ALGORITHMIC TECHNIQUES FOR THE POLYMER SCIENCES

ALGORITHMIC TECHNIQUES FOR THE POLYMER SCIENCES

Edited by
Bradley S. Tice, PhD

Apple Academic Press
TORONTO NEW JERSEY

Apple Academic Press Inc.
3333 Mistwell Crescent
Oakville, ON L6L 0A2
Canada

Apple Academic Press Inc.
9 Spinnaker Way
Waretown, NJ 08758
USA

Exclusive worldwide distribution by CRC Press, a member of Taylor & Francis Group

Printed in the United States of America on acid-free paper

International Standard Book Number-13: 978-1-926895-39-0 (Hardcover)

Library of Congress Control Number: 2014944888

Library and Archives Canada Cataloguing in Publication

Tice, Bradley S. (Bradley Scott), author, editor
Algorithmic techniques for the polymer sciences/Bradley S. Tice, PhD.

Includes bibliographical references and index.
ISBN 978-1-926895-39-0 (bound)
1. Polymers. 2. Polymers--Compression testing. 3. Algorithms. I. Title.

TA455.P58T53 2014   620.1'92   C2014-904860-2

Apple Academic Press also publishes its books in a variety of electronic formats. Some content that appears in print may not be available in electronic format. For information about Apple Academic Press products, visit our website at **www.appleacademicpress.com** and the CRC Press website at **www.crcpress.com**

ABOUT THE EDITOR

Bradley S. Tice, PhD

Dr. Bradley Tice is Institute Professor of Chemistry at Advanced Human Design, located in the Central Valley of Northern California, USA. Advance Human Design is a research and development company with a focus on telecommunications, computing, and materials science. Dr. Tice is a Fellow of The Royal Statistical Society, a member of The Royal Pharmaceutical Society, and a Fellow of the British Computer Society. Dr. Tice is also CEO of Tice Pharmaceuticals that is located in San Jose, California U.S.A.

PREFACE

The monograph addresses the use of algorithmic complexity to perform compression on polymer strings to reduce the redundant quality while keeping the numerical quality intact. A description of the types of polymers and their uses is followed by a chapter on various types of compression systems that can be used to compress polymer chains into manageable units. The work is intended for graduate and post-graduate university students in the physical sciences and engineering.

Dr. Bradley S. Tice, FRSS & FBCS

CONTENTS

INTRODUCTION

The monograph examines algorithmic compression techniques for use on polymer chains. The author has previous publications on the area of algorithmic complexity and has addressed their use to polymers in this monograph (Tice, 2009 and 2010).

The monograph is arranged into chapters that address polymers, chemical processes, algorithmic complexity, and then applied aspects of algorithmic complexity to polymer chains. Each chapter is a self-contained section that addresses that particular topic and lends itself well to understanding the application and theory of polymer compression techniques.

The monograph addresses large chain and multiple sequence polymers that are currently found in daunting amounts in "Big Data", and a chapter and extensive appendix sections are added to address the 'real world' problems of massive data sets of polymer information.

REVIEW OF THE LITERATURE

The literature for this dissertation is concise with Mason (1953), Brandman, Ferrell, Li and Meyer (2005), Barnholdt (2005), and Tice (1996) being primary sources. The remaining literature represents secondary sources of research. I have included my further work with signal flow diagrams in Appendix A of this dissertation. The literature is, in essence, the utilization of engineering techniques to natural, biological, and processes. The resulting list of literature becomes neat and tidy as a result of the direct aspects of this academic work. Signal flow diagrams are graphs that were invented by Mason in 1953 (Mason, 1953).The signal flow diagram is a directed graph that may have cycles and loops present that represent feedback in the system.

This type of graph is also considered a network (Busacker and Saaty, 1965). While signal flow diagrams are the domain of the engineering disciplines, a growing need for such graphing methods is developing in the physical and natural sciences, especially in the field of systems biology (Barnholdt, 2005). The dissertation will take current research data from the field of systems biology and model interlinked fast and slow-positive feedback loops in producing reliable signaling decisions for cells (Brandman, Ferrell, Li, and Meyer, 2005 and Bornholdt, 2005).

CHAPTER 1

POLYMERS

BRADLEY S. TICE

INTRODUCTION

A polymer is a chemical material that consists of repeating structural components which are formed through the process of polymerization (Wikipedia, "Polymer", 2013). The word polymer comes from the Greek words "poly" to mean "many" and "meros" to mean "parts" (Borchardt, 1997).

The shapes of polymer chains, molecules linked together to form a sequence of connected molecules, has several types of geometries beyond the linear polymer chains (Hiemenz and Lodge, 2007). Branched and cross linked polymer chains are common and arise from the "backbone" of a linear molecule (Hiemenz and Lodge, 2007). The amount of such polymer branching structure is a branching upon branching of a molecule will result in a network type of geometry that is termed cross linked (Hiemenz and Lodge, 2007). Some multi-branched molecules have discrete units and are termed hyper-branched polymers and other multi-branched polymers known as dendrimers, or tree-like, molecules (Hiemenz and Lodge, 2007).

Co-polymers are repeating units of polymers that have more than one type of repeating polymer unit and a polymer chain that has only a single type of repeating polymer unit is termed homo-polymers (Hiemenz and Lodge, 2007). So, a co-polymer is a series of monomers that repeat in a chain and are bounded by each of their original monomer states (Hiemenz and Lodge, 2007). The tertiary structure of polymer is the "overall" shape of a molecule and formal polymer nomenclature uses the structure of the monomer or repeat unit as a system of identification by the International Union of Pure and Applied Chemistry (IUPAC) (Hiemenz and Lodge, 2007).

Monomers and repeat units of monomers are the primary descriptive quality of polymers and are categorized in the nomenclature according to the type of structures involved in the monomers (Wikipedia, "Polymer", 2013). Single type of repeating monomers are known as homopolymers, while the mixtures of repeat monomers are known as co-polymers (Wikipedia, "Polymer", 2013).

The arrangement of monomers in co-polymers is as follows (Wikipedia, "Polymer". 2013):

ALTERNATING CO-POLYMERS:

-A-A-A-A-A-A-A-A-A-A-A-A-

PERIODIC CO-POLYMERS:

-A-B-A-B-A-B-A-B-A-B-A-B-

STATISTICAL CO-POLYMERS:

-A-B-B-B-A-B-A-B-A-A-A-A-

BLOCK CO-POLYMERS:

-B-B-B-B-B-B-A-A-A-A-A-A-

GRAFT CO-POLYMERS:

```
         - A - A - A - A - A - A - A - A - A - A - A - A
                       |              |
- B - B - B - B - B                   B - B - B - B - B -
```

Polymer architecture is the microstructure that develops the way the polymers branching points lead to a variation from a linear polymer chain (Wikipedia, 2013, "Polymer" 7). A branched polymer molecule is composed of a main chain of polymers and one or more branched sub-chains that form a geometric list of branched polymer forms as follows (Wikipedia, 2013, "Polymer": 7-8).

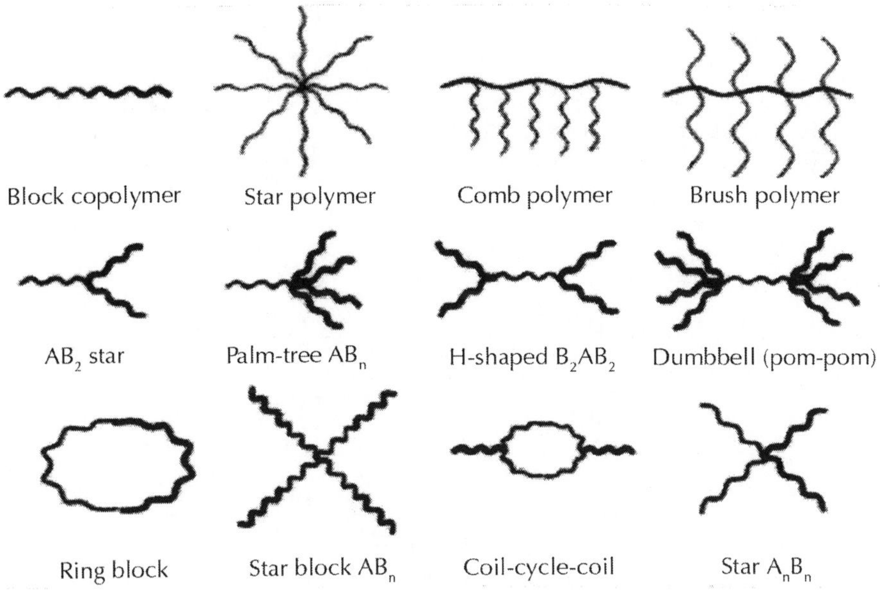

VARIOUS POLYMER ARCHITECTURES.

The physical properties of a polymer are dependant on the size and length of a polymer chain Wikipedia, 2013, "Polymer": 7-8). The polymerization is the length of the chain of polymers and thephysical size is also measured as a molecular weight.

Polymer	Important Properties	Uses
Low-density polyethylene (LOPE)	good toughness and pliability, transparent in films, excellent electrical insulator, resistant to chemicals	films used in food wrapping and other applications, drapes, tablecloths, electrical wire and cable insulator, squeeze bottles

TABLE *(Continued)*

High-density polyethylene (HOPE)	higher crystallinity, softening temperarure, hardness, and tensile strength than LOPE	bottles (especially for liquid laundry detergent), coatings, pipes, wire and cable insulation
Polypropylene	relatively low density; high tensile strength, stiffness, and hardness	carpeting, injection-molded pans for appliances, photocopiers, and other machines
Polyvinyl chloride	resistant to fire, moisture, and many chemicals, degraded by heat and ultraviolet light	food containers, floor coverings, films, rainwear, coatings for electrical cables and wires
Polystyrene	transparent and colorless, easily colored by pigments, good resistance to chemicals, good electrical insulator	foam packing material, plastic pans, housewares, toys, packaging
Polytetrafluoroethylene (Teflon@ polymer)	chemically inen, low friction propcnies	coatings for cookware, insulation for wires, cables, motors, nonlubricated bearings
Polymethyl methacrylate	transparent, colorless, high impact strength, poor abrasion resistance	applications in which light transmission is needed: signal light lenses, signs, eyeglass lenses

TABLE *(Continued)*

Styrene-butadiene rubber	equal or better physical properties compared with natural rubber except for heat resistance and resilience	tire treads for cars (but not trucks, where natural rubber is used due to its better heat resistance and resilience)
Polyvinyl acetate	water-sensitive, good adhesion	water-based latex paints, to produce polyvinyl alcohol, low-moleculearweight polymer in chewing gum
Polyvinyl alcohol	water-soluble, fair adhesion	water-thickening agent, packaging film
Polyvinyl butyral	good adhesion to glass, tough, transparent, resistant to ultraviolet degradation	an inner layer in automotive safety glass (windshields)

The products of polymers started in 1839 with Charles Goodyear with the mixing of sulfur and natural rubber to produce a useful product (Borchardt, 1997:1230). Same popular uses for polymers are industry manufactured synthetic fiber produced by melted polymers and found in the following common fibers (Borchardt, 1997: 1234).

Polymer Type	Comments	Typical Uses
Nylon	amide group repeat units	Carpeting, upholstery, clothing, tire cord
	Acrylonitrile units constitute 90% or more of the polymer weight	Carpeting drapes, blankets, sweaters, other articles of clothing

TABLE *(Continued)*

Polyester	Copolymer of a diacid and a diol (molecule with two hydroxyl groups)	Permanent press clothes, underwear, 2-liter soda bottles
Spandex	Polyurethane	Athletic clothes, girdles, bras
Rayon	Made from cellulose	Clothing, blankets, curtains, tire cord

The examples used in this monograph are basic linear models of polymer chains to give emphases on their application to the algorithmic complexity program.

Products of polymers started in 1839 with Charles Goodyear with the mixing of sulfur and natural rubber to produce a useful product (Borchardt, 1997:1230). Same popular uses for polymers are industry manufactured synthetic fiber produced by melted polymers and found in the following common fibers (Borchardt, 1997: 1234).

Polymer Type	Comments	Typical Uses
Nylon	Amide group repeat units	Carpeting, upholstery, clothing, tire cord
	Acrylonitrile units constitute 90% or more of the polymer weight	Carpeting drapes, blankets, sweaters, other articles of clothing
Polyester	Copolymer of a diacid and a diol (molecule with two hydroxyl groups)	Permanent press clothes, underwear, 2-liter soda bottles
Spandex	Polyurethane	Athletic clothes, girdles, bras
Rayon	Made from cellulose	Clothing, blankets, curtains, tire cord

The examples used in this monograph are basic linear models of polymer chains to give emphases on their application to the algorithmic complexity program.

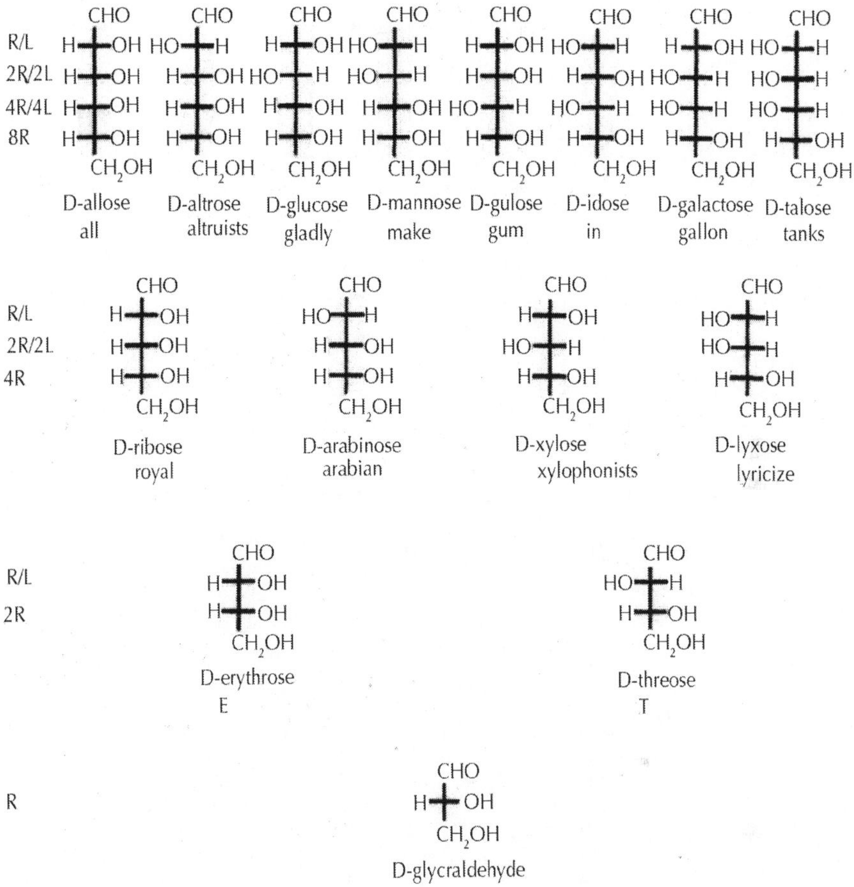

The *D*- family tree of the ketoses.

CHAPTER 2

COMPRESSION OF DATA

BRADLEY S. TICE

INTRODUCTION

A mathematical notion of compression can be found in the fields of computer science and information theory (Wikipedia, "Data Compression" 2013). The use of fewer bits of information to represent the original code structure of information, or data, is termed either data compression, source coding or bit-rate reduction and is either a lossless or lossy compression format (Wikipedia, 'Data Compression', 2013). The lossless compression is the reduction of statistically redundant information with no loss to the original information content (Wikipedia, "Data Compression", 2013). Lossy compression is the removal of unnecessary information from the original source, hence the term "source" coding, with the resulting loss of the amount of original information or data (Wikipedia, 'Data Compression', 2013).

The symmetrical data compression is when the time to compress is the same as decompression and asymmetrical data compression is when compression and decompression times vary (Wikipedia, "Data Compression Symmetry", 2013).

Universal code data compression is a prefix code that transposes positive integers within binary code words that are monotonic with relation to statistically probable distribution of lengths an optimal code would have produced (Wikipedia, "Universal Code", 2013). On average prefix codes are assigned longer code words to larger integers, but are not used for precisely known statistical distribution and no universal code is known to be optimal for probability distribution (Wikipedia, "Universal Code", 2013). A universal code is not a universal source code as no fixed prefix code is used and both Huffman coding and arithmetic encoding are better at compression than universal coding, unless the exact probability of the message is not known and then a universal code is needed for compression (Wikipedia, "Universal Code", 2013).

Huffman codes where created in 1951 by David A. Huffman, a Ph.D. student at Massachusetts Institute of Technology (MIT), and is an entropy algorithm for encoding lossless data compression (Wikipedia, "Huffman Coding", 2013).

Huffman codes use a specific type of representing and choosing each symbol that is designated a prefix free code so that any bit string represent a specific symbol is never the prefix of another symbol (Wikipedia, "Huffman Coding", 2013). There are many variations of Huffman codes (Wikipedia, "Huffman Codes", 2013).

The arithmetic coding, on average, have better compression values because of arbitrary number of symbols that can be combined formore efficient coding and function better when adapting to real world input statistics (Wikipedia, "Huffman Coding", 2013).

The entropy encoding data compression is independent of the specific characteristics of the medium used and is a lossless compression code (Wikipedia, "Entropy Encoding", 2013). By assigning a unique prefix free code to each specific symbol in the input and replace each fixed length input symbol with a variable length prefix free output code word for compression (Wikipedia, "Entropy Encoding", 2013). The most common symbols use the shortest codes for compression (Wikipedia, "Entropy Encoding", 2013).

DECOMPRESSION

Decompression is the state of returning a compressed sequence to its original size and sequential origins. There is exactness to this process of resurrecting a compressed sequential string in that both the accurate placement and type of symbol must occur to be the 100% realization of the original pre-compressed state of the sequential string.

A compression of that binary sequential string, [11001100], would be as follows, [1010], reducing the total symbol quantity by half, from 8 symbols to 4 symbols, thus producing a successful compression of the original 8 symbol sequential string.

A decompression of the compressed sequential string will produce the following key decompression measures:

That the initial, and following symbols, to the last, or terminal, symbol match the original sequential strings symbol placement and position within the totality of the original sequential string.

That there are no more, and no less, than the exact number of symbols present in the pre-compressed original state sequential string.

An accurate and precise decompression is mandatory for a useful application of this algorithmic complexity technique.

In simplest terms compression is a process of 'reducing' redundant data from an object. This redundant data is an original part of the object and so it must be said that it is an original feature of that object, even though it may seem excessive to the original design of that object. An example from genetics has redundant genetic instructions, hence the term 'junk' DNA, as being an important coding system for supporting healthy development of the organism (Princeton University, 2010: 1–3).

Take an object that is made up of a simple alphabet of four letters: ABCD and that alphabet is used in a linear manner to function as a symbolic relationship to each other to form a strict hierarchy of spatial interaction in that the symbol A is always in the initial position always followed by the symbol B that is always followed by C with the Symbol D always in the last position. The grammar of this ABCD alphabet would always be as follows: ABCD, with no variation permitted by the rules of concatenation by each symbol of that four letter alphabet. If this ABCD alphabet was allowed to reproduce itself it would look like this arrangement: ABCDABCDABCDABCD, with the original alphabet being the initial four symbols; ABCD, and the duplication of three more sets of the original four symbol alphabet added in the suffix position.

A compressed version of this ABCD 'sentence', a group of words, symbols, unified to form a sentence, in this case the sentence being ABCDABCDABCDABCD, from the word ABCD. A compressed ABCD sentence would be [ABCD] plus ABCD added three times to the original ABCD word. The 4 symbol word was expanded into a 16 word sentence that could be compressed into a four symbol alphabet. By de-compressing the compressed ABCD sentence into a full 16 symbol ABCD sentence. Compression should be seen as a 'reduction' of the original objects duplicate features, but not the reconfiguring of those original object total features.

Compression and de-compression should be seen as mirror properties of the same object, nothing is lost, just hidden. A binary sequential string such as [11001100] is composed of two 1's in the initial position followed by two 0's and then two 1's with two 0's in the final two positions of this linear sequential string.

CHAPTER 3

NATURAL LANGUAGE COMPRESSION

BRADLEY S. TICE

INTRODUCTION

Natural languages are those spoken by human beings and will be examined in this study as a written language rather than the traditional verbal or spoken and language.

The texting, the faddish uses of telephones, landlines, mobile devices, and cell phones, that use, in the english language system, 26 letters of the alphabet and 10 numerical symbols, an alphanumeric system, to send a message from one source to another source *via* a transmission system (Wikipedia, "Texting Messaging", 2013).

The act of compression is a natural act of parsimony of a sentence to reduce the amount of characters used without the loss of the "content" of the message coded into this abbreviated form. Because the vowels of the english language, a, e, i, o, and u are the connecting units of consonants they form the "significant" junction of words, especially prefix, suffix, and non-traditionally, infix positions in the english language and by themselves, something english consonants do not normally do, become singularities of form and represent themselves accordingly.

Texting of the English language uses the "dropping" of a vowel within english word to save on space. The following example of such an english sentence is—"Please pick me up at ten" can be shortened by dropping the "ea" in Please, the "ic" in pick, and the letter "ten" can be changed to the numerical 10 resulting in the following texting "shorthand"—"Plse pk me up at 10" that has shortened the original 19 character message to the texting message's 14 character length.

Poetry is the most common form of language compression, both from a physical perspective, usually shorter than other forms of storytelling such as short stories, novels etc., and grammatically, made up words and non-traditional or a 'slang' vernacular forms of speech that have a minimal word formation. Where poetry is also compressed is in the semantic value of the poem, in that fewer words are used to produce a more expanded notion of meaning, or meanings, than a traditional use of words and sentences. Rewriting or editing a written work of literature is a form of compressing, in that how a word or sentence is composed can be duplicated in a shorter form by the addition of changing of words and sentences within a completed literary work. It has long been believed that short words are used more than long words due to George Zipf's 1930's work on frequency of words used by a peaker (MIT, 2011: 1-3). New research finds that the lengths of words are the amount of information the word contains, not the requency of the use of that word (MIT, 2011: 1).

Even words that are misspelled in a sentence are 'automatically' corrected and interpreted as being a; normal spelling, but there is a limit the number and types of 'jumbled' letters in a sentence for a native speaker to adjust to a readable pattern (ESRC, 2013: 1-3).

Number words have long had a place in man's history especially the number two; 2, that has the suggestive influence of being duel and mirror like in that human beings are bilateral and have two legs and two arms, two eyes, etc and that common objects come in pairs (Menninger, 1969:12). The act of 'counting' as described by Otto Jesperson, leads to objects as "more than one" without those objects being identical and that plurality presupposes difference as long as such differences have a conceptual common relationship at the core, such as a pair' and an 'apple, but not a 'brick and a castle'(Jesperson,1924/1968:188).

The noted linguist, Benjamin Lee Whorf (1897–1941) found that the Native American Indian language of the Hope Tribes use of verbs have a more precise and variant time and space correlation that European languages in that a physical object, when placed on a specific point; space, say a building on a fixed point of land, must be qualified as 'when'; time, the building was on that spot of land, and as such is a more perfect language for communicating space and time (Sebeok, 1966:580).

Whorf also study the Hopi noun system and found the descriptive qualities to have a large measure of variation. The concept of 'rain' is not just 'water falling from the sky', but is a descriptive quality of the 'properties' of the type of 'rain' being described such as: wet rain falling from sky, wet rain that runs in a river, frozen rain; snow, frozen rain pellets; hail, etc, and that each of these physical actions are described by a separate noun formation (Carroll, 1956:140).

Numbers and words have a long history with both strong linguistic and mathematical lineage and the expansion and utilization of chemical and algorithmic properties adding to this long and developing history of man and his description of the world around him.

CHAPTER 4

FORMAL LANGUAGE COMPRESSION

BRADLEY S. TICE

INTRODUCTION

The formal languages are computer languages and have their own rules and laws of formation, grammars, and connectivity, the structural qualities inherent in unions of parts of the whole. These are, by nature, linear and have the qualities of being mathematical in which they are represented as binary sequential strings.

The computers use formal languages that are limited to type of the input devices to input the information and the primary "limiter" to this information is the computer key board, a modified version of the "QWERTY" typewriter key board that was invented in 1878 with the popular success of the No. 2 Remington typewriter (Wikipedia, "QWERTY", 2013). The language of mathematics is system of both natural languages, say the English language, that is invested with technical terms and grammar conventions to a specialized symbolic notation system (Wikipedia, 2013, "Language of mathematics": 1-8). The vocabulary of mathematics is by the use of notations that have grown over the years as mathematical developments have occurred.

The grammar of mathematics is the specific use of a mathematical notation system, the rules that underlie 'how' and 'when' a particular notation is to be used with the framework of a mathematical discipline (Wikipedia, 2013, "Language of mathematics": 4). An example would be the English word formula one plus one equals two that can be mathematically notated as $1 + 1 = 2$. A hybrid of the two language systems, natural language and mathematical formalisms such as the word formula for one plus one equals two can be written as 1 + one equals 2.

The author has done research in the area of natural language and mathematics with a work addressing Kurt Godel's Continuum Theory use of word problems as proof of inconsistency in David Hilbert's axiomatic laws of mathematics. The author changed some of the words found in Godel's original paper that, in turn, changed the semantic nature of the sentences used to proof the axiomatic truths found in Godel's paper and by so doing changed the valuations used to determine an inconsistency (Tice, 2013).

Current research has found that a similarity in genetic codes and computer codes arises when the most frequently used of components of either a biological or computer system have the most 'descendants' (Brookhaven National Laboratory, 2013: 1).

Crucial parts of a genetic code in the metabolic process of 500 bacteria species was measured against the frequency of 200,000 packages of an open source software program, Linux, and that the more a code, either biological or computer, it is adopted more often and takes up a greater portion of the system (Brookhaven National Laboratory, 2013: 2).

The rules of a formal language have specific grammars that give clues to a message, or signal, being evaluated from outside the intended communications channel. In long series of passwords, coded words or properties of multiple words used to allow access to a secure message, that are based on words or phrases; sentences, have a distinct 'grammar' pattern that can be 'de-cyphered' more accurately than a more random sequence of symbols. The longer the sentence, the easier it is to find the frequency of common usage in that some aspects of grammar, such verbs and nouns are more prevalent that other aspects of a languages grammar pattern (Carnegie Mellon University, 2013: 1). Niels Jerne, in his Nobel Prize Lecture, compares the human body's immune system to 'inherent' traits of human language acquisition and that the 'seeds' of this genetic transference is in the 'DNA' of the human progenitors of the species (Jerne, 1984: 211-225).

This monograph will address carbohydrate molecules as sources of polymer chains, the 'chains of life', or DNA, are genetically encoded polymers containing life's language for continuing life. These genetic codes have set rules for functioning and become either viable or non-viable mutations when changes occur in the 'processes' of encoding or decoding a specific point within a genetic code.

While this monograph does not utilize genetic codes for either theoretical or applied aspects of algorithmic complexity programs, their application to both theoretical and applied genetics is for future publications and while some papers on theoretical genetics are included in the appendix section of this monograph, a major work on the subject must wait for future research papers and monographs.

CHAPTER 5

TYPES OF COMPRESSION PROGRAMS

BRADLEY S. TICE

INTRODUCTION

The entropy coding is used in computer science and information theory in the form of Huffman coding that produces lossless data compression (Wikipedia, "Huffman coding", 2013). Huffaman codes and "prefix" codes are synonymous as Huffman codes are ubiquitous in computing.

The type of compression techniques to be examined in this monograph are algorithmic compression techniques as developed by the author and address both whole segments of common, or like-natured, polymer units and specific, or truncated, and aspects of a linear sequential string or a random or non-random type.

A non-random binary sequential string is as follows:

[11001100110011]

A series of alternating binary characters sub-grouped into two like-natured characters that have a total length of 14 binary characters. A random binary sequential string is as follows:

[100011011000001111]

A pattern-less series of alternating binary characters, 18 characters in length those are not compressible by traditional measures of compression. A compressible random binary sequential string is introduced in the next chapter and was discovered by the author in 1998.

CHAPTER 6

ALGORITHMIC COMPRESSION

BRADLEY S. TICE

INTRODUCTION

Algorithmic complexity is a sub-field of Claude Shannon's Information Theory (1948) and is known by various names as algorithmic information theory, kolmogorov complexity, and algorithmic complexity to name a few such monikers and was developed by R. J. Solomonoff, A. N. Kolmogorov, and G. J. Chaitin in the 1960's (Chaitin, 1982).

The traditional definition of algorithmic compression is the degree of difficulty, or complexity, needed for an object, usually a binary sequential string, to calculate or "construct" itself (Chaitin, 1982). On a perceptual level this notion of complexity is associated with degrees of measurable "randomness" found in a binary sequential string and is considered a form of Martin-Lof randomness because of visible nature of the measure of randomness (Chaitin, 1982: 40).

This "visible", or perceptual, nature of non-randomness is found in the regularity of a binary system, either l's or 0's, in that a regular pattern of l's and 0's would look like—1010101010 or five l's followed each by five 0's in a regular pattern of alternating symbols for a total of ten characters. A "random", or irregular pattern, would look as follows—1001110110 or a ten character total with a concatenated sub-grouping of [1] [00] [111] [0] [11] [0]. The significance of this finding is that a measurable nature of randomness could be described in statistics for the first time and set a defined boundary between randomness and non-randomness. This measure of randomness and non-randomness also had the properties of compressibility and non-compressibility of a binary sequential string.

The author discovered in 1998 a "compressible" random binary sequential string and stands as the most precise and accurate measure of randomness known in statistical physics (Tice, 2009 and 2012). A compressible random binary sequential string program utilizes a traditional binary random sequential string such as [001110110001110] that is 15 characters in length and sub-groups each like-natured sub-group of either 0's or 1's as follows:

[00] + [111] + [0] + [11] + [000] + [111] + [0]

A notation system is used behind each initial character to give a total number of like-natured characters in each sub-group while removing the remaining like-natured characters from each sub-group as follows:

0(2) + 1(3) + 0 + 1(2) + 0(3) + 1(3) + 0

The compressed state of the original random 15 character binarysequential string is as follows:

0101010

The compressed state is 7 characters in length from the original pre-compressed state of a 15 character random string.

Traditional literature on random binary sequential strings has this random 15 character binary sequential string as being unable to compress. The author's method of compression reduces and compresses the original 15 characters into less than half of the original total length.

The author has expanded the use of radix based number systems beyond the traditional binary, or radix 2 based number system, including two digit radix numbers; 10, 12 and 16, as well as radix 3, 4, ,5 and 8 based number systems (Tice, 2013). These larger radix based number systems can be more easily used with ternary and larger symbol systems that use more than a dual, or binary, system of characters. Analog systems are usually associated with more than two types of features, such as the four bases of the genetic code, that make it an ideal system for a radix 4 based number system.

While the symbols used for the characters are Arabic numbers, the use of [1] and [0] are to note contrasts of the [1] and the [0] Arabic symbols and not the quantitative numeration of a one [1] or a zero [0] number. The use of other symbols to represent the contrastive nature of the [1] and the [0] could be an alphabet that could use the letter [A] for [0] and the letter [B] for [1]. This minimalism is the reduce the symbol to a two natured contrast of black and white, night and day, up or down type of referencing that denoted the opposite nature of each symbol in use.

Linguistics uses many binary systems in giving a 'scientific' description of a language process. Phonology, the study of sound patterns of spoken human languages uses a contrastive feature of a phonological property of a specific sound per a single or two adjoined letters of an alphabet to assign a specific sound. When these letters make up a word, the letters in that word make up either a consonant or a vowel. The following phrase, a phrase is a spoken sentence, "The chemist is a scientist" can be written as a consonant and vowel representation of the alphabetical letters as sounds of that phrase. The phrase "The chemist is a scientist" can be written as "ccv ccvcvcc vc v ccvvccvcc" with the c = consonant and v = vowel. This is a binary representation of a consonant and vowel model of the phonological pattern of an English language phrase as written as an English language sentence.

Because the areas of study are of a finite length, the techniques employed with involve either the whole finite string of polymers or subsections of that whole finite string of polymer chains. The use of an algorithmic complexity program to 'compress' only a section of a finite length of a polymer string can be applied in the same manner as the whole string. Each subsection of a whole finite string is made of common components of that whole finite string.

An example would be the following random binary sequential string of a twenty character length:

[11000100000111000110]

If the desired compression area within this whole finite string was the unit of [0] that comprised five sequential [0] symbols then the seventh symbol, from the left initial side position, [0] and the sequential 4 [0's] following that seventh symbol are compressed a [0]x 5 as a notation that five sequential[0]symbols have been compressed and the can be represented as follows:

[110001{0}111000110]

Notice that brackets have been used to note the placement and symbol type of the 'compressed' 5 [0] section of the finite string.

The remaining 16 character length finite sequential binary string is reduced from the whole finite sting of the original 20 character string and reduced by four characters.

Decompression of the compressed section to its original placement and character type and results in an exact reproduction of the original whole finite string.

Algorithmic Complexity is a sub-field of Claude Shannon's *Information Theory* that was published in 1948 and has been adopted for use in chemistry in two monographs (Eckschlager and Danzer, 1994 and Eckschlager and Stepanek, 1979). Both Information Theory and Algorithmic Information Theory have been evaluated on 'esthetic' principles as published by Moles (1966) and Stiny and Gips (1978).

CHAPTER 7

CHEMICAL FORMULAS

BRADLEY S. TICE

INTRODUCTION

Chemical formulas are a way to express the number of atoms that make up a particular chemical compound using a notation system made up of a single line of chemical element symbols, numbers, and other symbols and are limited to a single typographic line (Wikipedia, "Chemical Formula", 2013).

Molecular formulas represent the number of each type of atom in a molecule of a molecular structure (Wikipedia, "ChemicalFormula", 2013). Empirical formulas are the letters and numbers indicating atomic proportional ratios of one type of atom to another type of atom (Wikipedia, "Chemical Formulas", 12013).

Chemical formulas can be written as follows (Cartage.org., "Structural Formulas", 2013):

Common Name: Formaldehyde
Molecular Formula: CH_2O
Lewis Formula:

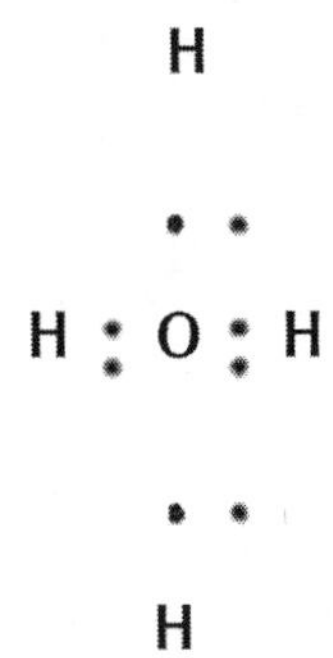

KEKULE FORMULA:

H

I

H - O - H

I

H

The importance of chemical nomenclature is that the 'language' of chemistry is by way of signs and symbols that denote a specific property or action within a chemical structure. This monograph will be using Fischer Projections of carbohydrates and linear notation systems for chemical formulas as examples for compression of polymer molecules. Herman Emil Fischer developed the Fischer projection in 1891 as a two dimensional representation of a three dimensional organic molecule usually a carbohydrate (Wikipedia, 2014,"Fischer Projection": 3). A Fischer Projection will give the impression that all horizontal bonds project toward the viewer and the vertical bonds project away from the viewer (Wikipedia, 2014, "Fischer Projection": 1). Fischer projection are used to depict carbohydrate molecules and give a differentiation between *L*- and *D*- molecules with *D*-sugar carbon units having hydrogen on the left side and hydroxyl on the right side of the carbon backbone (Wikipedia, 2014, "Fischer Projection": 1). L-sugar carbon units will have hydrogen on the right side and hydroxyl on the left (Wikipedia, 2014, "Fischer Projection": 1). Fischer Projections are used in biochemistry and organic chemistry to represent monosaccharaides and amino acids as well as other organic molecules (Wikipedia, 2014,"Fischer Projection": 2).

The use of existing chemical nomenclature is to facilitate the ease of applicability, both theoretical and applied, to the use of an algorithmic complexity program for the polymer sciences.

CHAPTER 8

FISCHER PROJECTION

BRADLEY S. TICE

A Fischer projection is a two-dimensional depiction of a threedimensional organic molecule by way of visual projection (Wikipedia, "Fischer Projection", 2013). Hermann Emil Fischer developed this graphing system in 1891 (Wikipedia, "Fischer Projection", 2013). The characteristics of Fischer projections are one or more stereogenic centers surrounded by four plain bonds aligned in a vertical fashion (Brecher, 2006). These Fischer projections were originally designed to depict carbohydrates (Brecher, 2006).

In Fischer projections all bonds are represented as horizontal or vertical lines and are the only representational configuration that are not in variant with respect to rotation of depiction (Wikipedia, "Fischer Projection", 2013 and Brecher, 2006).

FISCHER PROJECTION

A Fischer Projection is a two-dimensional depiction of a threedimensional organic molecule by way of visual projection (Wikipedia, "Fischer Projection", 2013: 1). Hermann Emil Fischer developed this graphing system in 1891 (Wikipedia, "Fischer Projection", 2013: 1). The characteristics of Fischer Projections are one or more stereogenic centers surrounded by four plain bonds aligned in a vertical fashion (Brecher, 2006: 1933) . These Fischer Projections were originally designed to depict carbohydrates (Brecher, 2006: 1933).

In Fischer Projections all bonds are represented as horizontal or vertical lines and are the only representational configuration that are not invariant with respect to rotation of depiction (Wikipedia, "Fischer Projection", 2013: 1 and Brecher, 2006: 1934).

Both *L*-sugars and *D*-sugars are the "mirror" image of the other with the *D*-sugars occurring naturally (Rensselaer Polytechnical Institute, 011: 2). Haworth Projections are similar to Fischer Projections and are used to depict sugars in ring form and Newman Projections are used to represent stereochemistry of alkanes (Wikipedia, 2014,"Fischer Projection": 3).

Examples of Fischer Projections (Rensselaer Polytechnical Institute, 2011: 1–2).

MONOSACCHARIDES

Aldoses (e.g., glucose) have an aldehyde at one end.

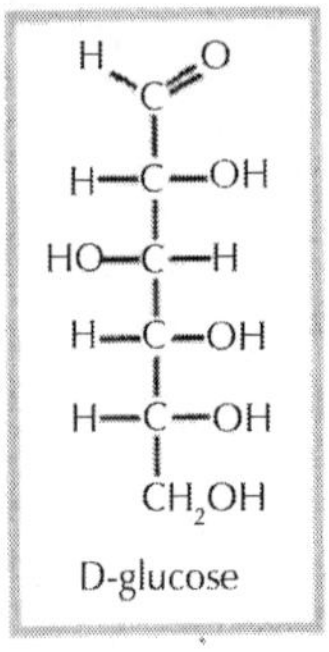

Ketoses (e.g., fructose) have a keto group, usually at C #2.

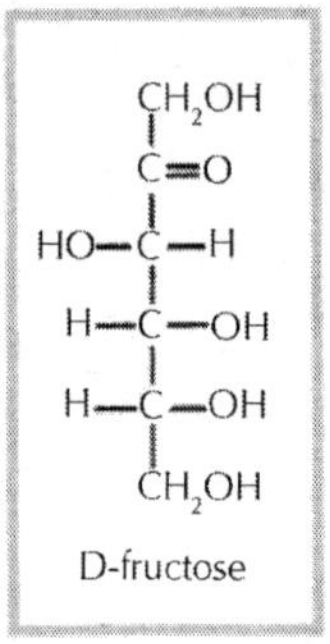

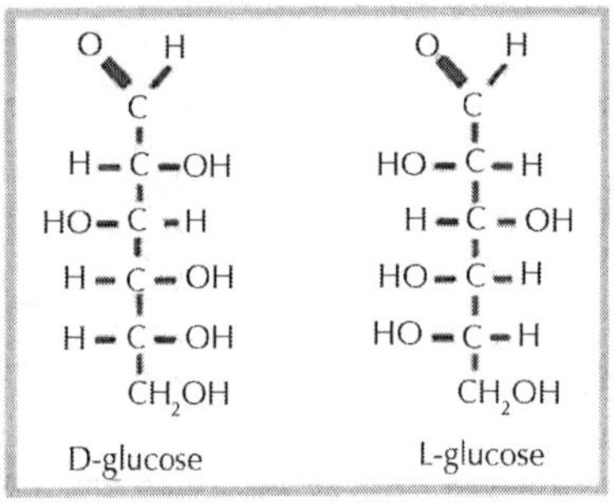

CHAPTER 9

COMPRESSION OF POLYMERS

BRADLEY S. TICE

INTRODUCTION

The notion of algorithmic complexity as it relates to a series of symbols in a linear pathway as found in the chemical depictions of Fisher projections is the "compression" of linear sequential similar symbols for reduction to a single symbol with a corresponding notation to denote the total number of characters of a similar type that were in the original chemical formula.

FISCHER PROJECTIONS OF CARBOHYDRATES

Monosaccharides are simple sugars with multiple hydroxyl groups (Molecular Biochemistry 1, 2013).

MONOSACCHARIDES (WIKIPEDIA, "2-CARB-3 CARB AND 2-CARB-4", 2013)

D-Glucose CHO-HCOH-HCOH-HCOH-HCOH-CH_2OH

A compressed version of D-Glucose would have the 4 HCOH units notated as HCOH (4) and would be written as follows:

D-Glucose (Compressed) CHO-HCOH (4)-CH_2OH

D-Xylose CHO-HCOH-HCOH-HCOH-CH_2OH

A compressed version of D-Xylose would have the 3 HCOH units notated as HCOH (3) and would be written as follows:

D-Xylose (Compressed) CHO-HCOH (3)-CH_2OH

D-Arabino-Hex-2-ulose CH_2OH-C=O-HOCH-HCOH-HCOH-CH_2OH

(D-Fructose)

A compressed version of *D*-Arabino-Hex-2-ulose would have the 2 HCOH units notated as HCOH (2) and would be written as follows:

D-Arabino-Hex-2-ulose CH_2OH-C=O-HOCH-HCOH (2)-CH_2OH
(Compressed)

D-Glycero-gulo-Heptose CHO-HCOH-HCOH-HCOH-HCOH-HCOH-CH_2OH

A compressed version of *D*-glycero-gulo-Heptose would have the 5 HCOH units notated as HCOH (5) and would be written as follows:

D-glycero-gulo-Heptose CHO-HCOH (5)-CH_2OH
(Compressed)

L-Arabinose CHO-HCOH-HOCH-HOCH-CH_2OH

A compressed version of *L*-Arabinose would have the 3 HCOH units notated as HCOH (3) and written as follows:

L-Arabinose CHO-HCOH (3)-CH_2OH
(Compressed)

L-glycero-D-manno-Heptose CHO-HOCH-HOCH-HCOH-HCOH-HOCH-CH_2OH

A compressed version of *L*-glycero-*D*-manno-Heptose would have the 2 HOCH and 2 HCOH units notated as HOCH (2) and HCOH (2) and written as follows:

L-glycero-*D*-manno-Heptose CHO-HOCH (2)-HCOH (2)-HOCH-CH_2OH
(Compressed)

CHAPTER 10

LINE NOTATION SYSTEMS AND COMPRESSION

BRADLEY S. TICE

INTRODUCTION

Line notation systems are used as a linear notation system to specify sub-structural patterns and structural patterns on chemical molecules (Wikipedia, "Smiles arbitrary target specification", 2013 and Wikipedia, "Simplified molecular-input line-entry system", 2013). Two systems are used by various chemical organizations, Simplified Molecular-Input Line-Entry System (SMILES) and Smiles Arbitrary Target Specification (SMARTS) (Wikipedia, "Smiles", 2013 and Wikipedia, "Simplified", 2013).

The compression of line notation systems is done at the point of repetition of a chemical symbol that makes it "redundant" in the formula's notation.

In the following notation for SMARTS a recursive SMART notation is used to combine acid oxygen and tetrazole nitrogen to define oxygen atoms that are anionic (Wikipedia, "SMARTS", 2013).

SMARTS Notation: [$ ([OH] [C, S, P] = 0), $ ([nH]1nnnc1)]

The obvious compression point is the three sequential "n" symbols [nnn] that can be compressed by either a numerical value for the "total" number of n's such as n3 or by notating with an underlining of a single "n" character to define a "quantity" of three "n" symbols, [nnn] = [$\underline{n}$]. Chromatic properties of notations, such a monochrome, black, or chromatic, colored, notation symbols can be used to signal "compression" points such as the example, [nnn], with the "n" symbol have a color other than the traditional black hue to signify three n's [nnn].

Using the SMILES notation system the following SMILES notation for cyclohexane is as follows (Wikipedia, "SMILES", 2013):

SMILES Notation: C1CCCCC1

The compression point is the five sequential C's [CCCCC] following the 1 in the formula. Again, a notational quantity of 5 C's can follow an initial C character to denote the number of C's symbols in the notation, C's = [CCCCC]. The use of an underlined C character could also be used to define the number of C's in the formula, C = [CCCCC].

CHAPTER 11

CURRENT TRENDS IN RESEARCH

BRADLEY S. TICE

INTRODUCTION

In reviewing the current literature on compression of polymers the following two types of areas appear—general polymers and bioprocess polymers genetics. General polymers are all non-genetic related polymers and bioprocess polymers are genetic oriented polymers for biological systems.

A sampling of the current literature on compression of genetic codes, deoxyribonucleic acid (DNA) and ribonucleic acid (RNA), has resulted in new developments in the analysis and use of genetic codes to science and engineering. Bioprocess polymers have advanced to the point of monographs being written in this area of science with the topics of DNA computing and algorithmic developmental processes in nature (Condon, et al., 2009).

GENERAL POLYMERS

Alagoz (2010) studied the effects of sequence partitioning on compression rates and found to have a greater compression rate on a sequence (Alagoz, 2010). The University of Wisconsin Madison (2010) notes that self-assembling polymer arrays improve data storage potential with the use of block co-polymers (University of Wisconsin-Madison, 2010).The Journal of the Amercian Chemcial Society (2010) that an organic ternary data storage device was developed using a three value system (J. ACS, 2010,). Evans, et al., (2001) used a symbol compression ratio for string compression and estimation of Komogorov complexity (Evans, et al., 2001).

BIOPOLYMERS

Svoboda (2010) DNA sequencing is done with a DNA “transistor” (Svoboda, 2010). In the *Physical Review of Letters* (1998) a notes the use of the entropic segmentation method to define the sequence compositional complexity of DNA (*Phy. Rev. Lett.*, 1998). Palmer (2011) uses a DNA computer to calculate square roots (Palmer, 2010). Pollack (2011) the huge amount of DNA sequencing is producing a large amounts of data (Pollack, 2011). Cherniavsky and Ladner (2004) study grammar-based compression of DNA sequences (Cherniavsky and Ladner, 2004). Translational genomics (TGen) (2010) uses an encoding sequencing method to compress genomic sequencing data (TGen, 2010). The University of Reading (2010) has created a synthetic form of DNA that functions as a linear sequence of letters for use in information technology (Phys.org, 2010). Inderscience (2009) notes the use of a computer data base to compress DNA sequences used in medical research (ScienceDaily, 2010). Condon, et al. (2010) has published a book on *Algorithmic Bioprocesses* that address organic compression of biopolymers (Condon, et al, 2010).

CHAPTER 12

BIG DATA

BRADLEY S. TICE

INTRODUCTION

Big data is the term that describes the ever growing amount of "data" that is being collected by various disciplines, most notably genetics. This "glut" of data has little form to make it a "content" and "contextual" value of "information" and is more a gigantic pattern of "excess" rather than an approachable part of an emerging whole.

The use of algorithmic complexity programs for compression of data streams and theoretical uses in modeling chemical and biological processes is a major factor in managing big data and giving a functional tool for addressing the major problem with large data pools—too much "data" and too little "information".

Also the type of "data" derived from new techniques has to be weighed against existing "information" on the resulting study. A current research project reported that a series of internal points, or nodes, in an organism have a great hierarchical value to the "operation" of that organism (Northwestern University, 2013). These "necessary" nodes can descript the whole process without "monitoring" the "actions" of the components of that organism (Northwestern University, 2013).

Similar findings about "internal" connectivity in organisms and the "behavior" of such organisms was clearly detailed the various "feedback" theory groups of the 1940's and 1950's, including Wieners "Cybernetics", and to a lesser degree, Shannon's information theory, not to mention Canon's concept of "homeostasis" in living systems that was done in the 1920's (Hayes, 2011). More of these "new" research programs are just extensions of older studies, such as "systems theory" such as Ludwig von Bertalanffy's *General System Theory* (1928) for biological processes (Wikipedia, "Systems Theory", 2013). Systems theory seems to be a subset of cybernetics theory in that cybernetics was engineering related and was a feedback oriented system that was based on external interacting feedback loops which produced finite results (Wikipedia, "Systems Theory", 2013). Bertalanffy's work is pre-dated by the Russian Alexander Bogdanov's three volume work that was published in Russia between 1912 and 1917 (Wikipedia, "Systems Theory", 2013).

It has included an unpublished manuscript titled *Modeling Complexity in Molecular Systems*: A Revised Edition (2013) that uses various techniques from the engineering field to feedback systems found in biological processes (Appendix L). These techniques, along with some advances in development in graph theory by the author, have added significant methodology to the analysis and modeling of aspects to big data.

By using such effective techniques as tools for managing and the analysis of big data can the situation be addressed in a productive manner without falling into the void of "unproductive" data collections (Ovide, 2013).

CHAPTER 13

MODELING COMPLEXITY IN MOLECULAR SYSTEMS: A REVISED EDITION

BRADLEY S. TICE

INTRODUCTION

The dissertation will model interlinked fast and slow positive feedback loops that represent reliable signal transmission to a cell's decision making process. The use of the signal flow diagram will be used to graph an ideal model of this system.

The growing need for effective modeling of complex systems in the biological sciences expands the traditional areas of graph theory from the physical sciences and engineering disciplines to that of systems biology. While the future of modeling biological processes and systems will be automated, by computers, the need for human evaluations of such processes will still fall withinthe sphere of human perceptions of those processes (Muggleton,2006).Hence, the need for clear and accurate graphing methods.The author has previous experience with flow graphs and has found that the application of such graphing methods to the natural sciences the ideal modeling tool for representations of systems and sub-systems (Tice, 1997a, 1997b, 1997c, 1997d, 1997e, and 1998).

The graphing method to be used in this dissertation will be the signal flow diagram, also called the signal flow graph, that was developed by Mason in 1953 (Mason, 1953). Because the signal flow diagram incorporates the use of cycles and loops to represent a feedback system, they present an ideal graphingmethod to represent the "product" flow of a system, in this case, a biological system. An example of current biological research data will be modeled using an interlinked fast and slow positive feedback loops to represent reliable signaling transmission for cell decisions (Brandman, Ferrell, Li, and Mayer, 2005, and Bornholdt, 2005). The model of the system data will use the signal flow diagram.

THE FLOW DIAGRAM

The signal flow diagram is a flow diagram that was developed by Mason in 1953 (Mason, 1953). The signal flowdiagram is a directed graph, also known as a digraph, that is a collection of elements known as vertices that when collected in ordered pairs, or arcs, have a direction (Faudree, 1987). The following is taken from mychemistry dissertation(1996) that describes a signal flow diagram (Tice, 2001).

SIGNAL FLOW DIAGRAM

The use of signal flow diagramsare common in fields suchasengineeringanda practicaluseofthemcanbe made in the field of pharmacology. The main reason for the use of signal flow diagrams over other diagram systems, formal or block diagrams are that they are easy to use and permit a solution practically upon visual inspection (Shinners, 1964). Signal flow diagrams can solve complex linear, multi-loop systems in lesstime than either block diagrams or equations (Macmillian, Higgins, and Naslin, 1964). A signal flow graph is a topological representation of a set of linear equations as represented by the following equation:

$$y_i = \sum_{j}^{n} a_{ij} x_j, \quad i = 1, \ldots\ldots, n \tag{1}$$

Branches and nodes are used to represent asetof equations in a signal flow graph. Each node represents a variable in the system, like node i represents variable y in Equation 1.Branches represent the different variables such asbranchij relatesvariableyitoyjwherethe branch originatesat node i and terminatesat node j in Equation 1 (Shinners, 1964).

The following set of linear equations is represented inthe signal flow graph in Figure 1 (Shinners, 1964).

y2 = ay, +by2 + cy4
y3 = dy2
y4 = ey1 + fy3
y5 = +gy3 +hy4

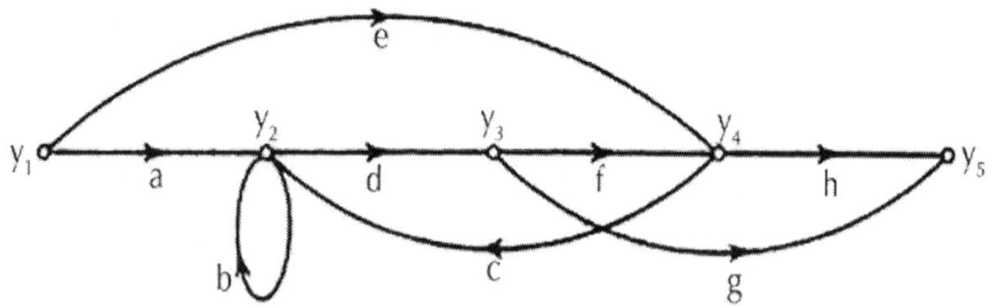

FIGURE 1 Signalflowdiagram.

It is necessary now to define the terms as represented by the signal flow diagram in Figure 1 (Shinners, 1964:28).

- The Source is a node having only outgoing branches, asy1 in Figure 1.
- The Sink is a node having only incoming branches, as y5 in Figure 1.
- The Path is a group of connected branches having the same sense of direction. These are he, adfh, and b in Figure 1.
- The Forward Paths are paths which originate from a source and terminateat a sink along which no node is encountered more than once, as are eh, adg, and adfh in Figure 1.
- The Path Gain is the product of the coefficient associated with the branches along the path.
- The Feedback Loop is a path originating from a node and terminatingatthesa-menode. In addition, a node cannot be encountered more than once. They are b and dfc in Figure 1.
- The Loop Gain is the product of the coefficients associated with the branches forming a feedback loop.

By using a signal flow diagram to represent the tables associated with pharmacological testing,drug delivery, behavior, dosage, and time intervals can all be graphed for ease of representation of these complex systems.

BINARY SYSTEMS IN BIOLOGY

The notion of a binary system is simple, whether it is off or it is on. It cannot be both, nor can it be a gradual quality of either. Black and white with no grey. In a biological system the process of communication, the science of signal transmission of information can be modeled into a binary system of two states, but with the added notion of duration, or time, it takes that signal to be transmitted as a variable.

Previous use of binary arithmetic to questions of biology can betraced back to Shannon's Doctorial work at M7T (1940) (Shannon,1940). The use of other disciplines "resources" is not uncommon today as an example Niels K. Jerne upon receiving his 1984 Nobel Prize in Medicine, gave a Noble lecture titled *The Generative Grammar of the Immune System* that seems to bind the study of linguistics with that of processes within the Luna system (Wright, 1988).

INTERLINKED FAST AND SLOW POSITIVE FEEDBACK

An example of interlinked fast and slow positive feedback is taken from a paper by Brandman, Ferrell, Li, and Meyer (2005).

In the paper biological systems, the case cell communication, is organized into a binary, on or off, system that uses positive feedback as the central method of communication (Brandman, Ferrel, Li, and Mayer, 2005). Multiple positive feedback loops can be composed of both fast and slow rates, the duration, it takes the signal to transmit acommunication, and that many of these multiple positive feedback loops are interlinked (Brandman, Ferrell, Li, and Meyer, 2005).

MODELING SYSTEMS BIOLOGY

The most important element to keep in describing a system or process is the integrity must be kept intact of that system or process. Because graphing is the most "illustrative" use of modeling of the descriptive devices to be used to describe a system or process, the need for an "ideal" or "most accurate" graph should be maintained to preserve the nature of the process being described. I have taken an examp1e of interlinked fast and slow positive feedback loops from a paper by Brandman, Ferrell, Li, and Meyer (2005) that describes reliable cell decisions in a schematic manner that makes for an ideal model for the signal flow diagram (Brandman, Ferrell, Li, and Meyer, 2005).

The focus of this example is not to examine the research in this paper but to use the signal flow diagram in place of the schematic model used in the paper. All explanations taken from the paper are to enhance the reason for the presentation of the data in a diagrammatic manner. This dissertation will assume theresearch from this paper is accurate and viable.

The focus will be on Brandman, Ferrell, Li, and Meyer's paper (2005) titled *Interlinked Fast and Slow Positive Feedback Loops Drive Reliable Cell Decisions* (Brandman, Ferrell, Li, and Meyer, 2005). In this paper, I am selecting the schematicmodel for the establishment of polarity in budding yeast cells (Brandman, Ferrell, Li, and Meyer, 2005). The paper addresses the presence of multiple interlinked positive loops asa question of performance as an advantage of the multi-loop design (Brandman, farrell, Li, and Meyer, 2005).

In using previous studies, Brandman, et al., concluded that the slow positive feedback loop was crucial for stability of the polarized "on" state, and the fast loop was crucial for the speed of the transmission between the unpolarized "off" state and polarized "on" state (Brandman, Ferrell, Li, and Meyer, 2005). The paper concludes that linked slow and fast positiveloop systems have advantages over single loop and dual looped systems such as independent timing of activation and deactivation times (Brandman, Farrell, Li, and Meyer, 2005).

The following is the positive feedback loop data from Table 1 in Brandman, et al. paper for budding yeast polarization (Brandman, Ferrall, Li, and Meyer, 2005). The information flow is from the left to right in both lines of data.

TABLE 1.

Cdc42–Cdc24–Cdc42
Cdc42–actin–Cdc42

The following is a copy of Figure 1 example A from Brandman, et al., paper that is a schematic representation of the data found in Table 1 of the polarization of budding yeast cells from the same paper (Brandman, Ferrell, Li, and Meyer, 2005).

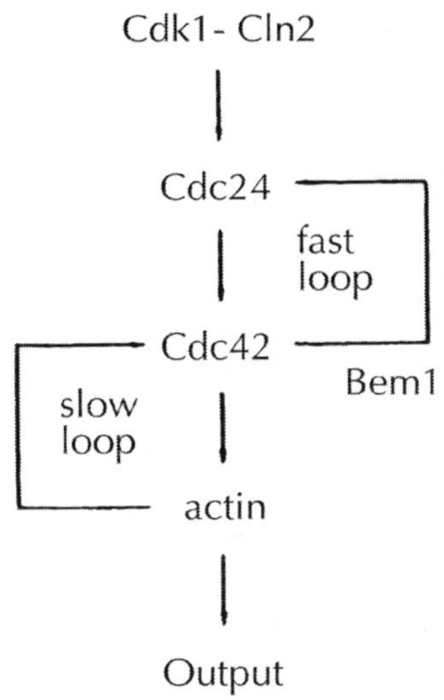

FIGURE 1 Example A.

In replacing the original schematic found in Brandman, et al., paper with a signal flow diagram the following will result as seen in Figure 2.

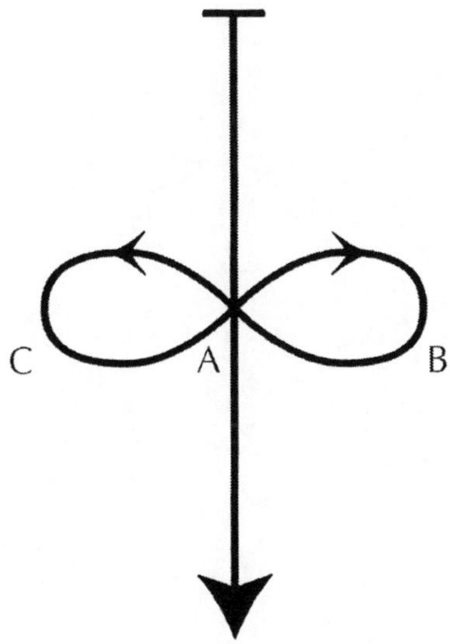

FIGURE 2 Signal flow diagram.

KEY TO FIQURE 2

Cdc42 = A
Cdc24 = B (Fast Loop)
Actin = C (Slow Loop)

COMMENTS ON FIGURE 2

1. Notice that the A node, Cdc42, is a common junction for both the Cdc24 and Actin feedback loops.The interlinking point of the system.
2. Loop B represents Cdc 24 and is the fast loop.
3. Loop C represents Actin and is the slow loop.

A more perspicuous manner for representing fast and slow aspects to the flow diagram can be done using chromatic variations of the traditional monochromatic signal flow diagram (Tice, 1996).By using a specific color for each loop, say red for the fast loop and green for the slow loop, a quick and easy solution presents itself upon inspection of the graph. Such a diagram is reproduced in Figure 3.

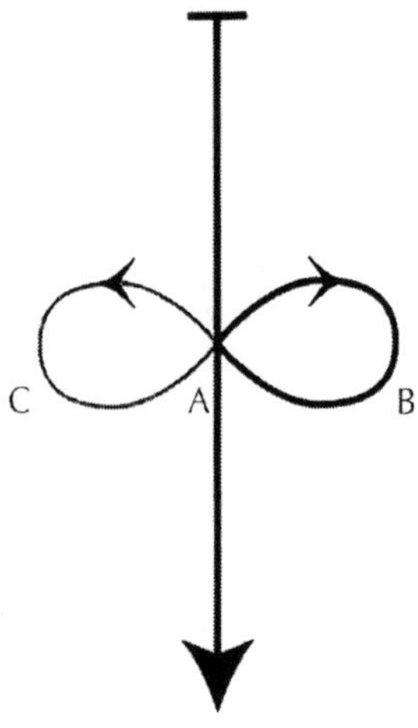

FIGURE 3.

KEY TO FIGURE3: Signal flow diagram.

Cdc42 =A
Cdc24 = B (Fast Loop) (Red Loop)

Actin = C (Slow loop) (Green Loop)

In comparing both the original schematic from the paper and the signal flow diagram version of the same data the following could be inferred.

1. While the original schematic was accurate and well realized, such notational qualities as direction of the cycle of feedback was vague and that all processes needed labeling, all points needed to be "spelled out" even though the process was a diagrammatic representationof the data flow presented in Table1, from the same page as Figure 1 from the same paper (Brandman, Ferrell, Li, and Meyer, 2005).
2. The signal flow diagram was a much simpler representation of the data flow from Table 1 and was made even more clear by the use of color. The Key to the signal flow diagram, the index orexplanatory part of the signal flow diagram, was also very simple and cogent.

In conclusion the signal flow diagram makes for a powerful replacement from the original schematic used in the paper (2005) and gives credence to the notion that correct representations of data are not just "good" or "bad" representations but also degrees of "viability"; which, at best, satisfies therepresentation, and that is 'ideal'; which being the best example of the representation.

SUMMARY

The signal flow diagram provided an ideal model of interlinked fast and slow positive feedback loops in representing reliable signal transmission of a cell's decision-makingprocess. Not only was it superior to the schematic representations in the original paper, but also underscores the real need for correct graphing methods when applying such methods from one field to another field.

With the advent of huge amounts of biological data being stored and interpreted, the real need for accurate and perceptible data, let alone information about the data, makes the need for presenting the data that is more important. Future research in this area could incorporate chromatic aspects of the signal flow diagram to enhance the information content and perspicuousnessof the biological systems being modeled.

CHAPTER 14

FEEDBACK SYSTEMS FOR NONTRADITIONAL MEDICINES: A CASE FOR THE SIGNAL FLOW DIAGRAM

BRADLEY S. TICE

INTRODUCTION

The signal flow diagram is a graphic method used to represent complex data that is found in the field of biology and hence the field of medicine. The signal flow diagram is analyzed against a table of data and a flow chart of data and evaluated on the clarity and simplicity of imparting this information. The data modeled is from previous clinical studies and nontraditional medicine from Africa, China, and South America. This report is a development from previous presentations of the signal flow diagram. The signal flow diagram has the conventional form of a circuit diagram, comprising connected branches and nodes. It differs from a circuit diagram in two important ways. First, the nodes are the points where the signals appear and where they are summated, and second, the branches are oriented and their orientation forms a single channel for the travel of the signal.

Mason introduced the signal flow diagram in 1953 as a process for analysis of linear systems, and as a graphic method of representation of a set of linear algebraic equations. When the equations represent a physical system, the graph depicts the flow of signals from one point of the system to another. In 1956 Mason would develop a gain formula that would give a transfer function of a linear system. The use of signal flow diagrams is common in fields such as engineering, and a practical use of them can be made in the field of biology. The main reason for the use of signal flow diagrams over other diagram systems are that they are easy to use and permit a solution practically upon visual inspection. Signal flow diagrams can solve complex linear, multiloop systems in less time than either block diagrams or equations.

A signal flow graph is a topological representation of a set of linear equations as represented by the following equation:

$$Y; = La;jxj, i = 1,...,n \tag{1}$$

Branches and nodes are used to represent a set of equations in a signal flow graph. Each node represents a variable in the system, like node "I" represents variable "y" in Equation 1. Branches represent the different variables such as branch "ij" relates variable "yi" to "yj" where the branch originates at node "i" and terminates at node "j" in Equation 1. The following set of linear equations are represented in the signal flow graph in Figure 1.

y2 = ay +by2 + cy4 y3 = dy2
y4 = ey1 +fy3
y5 = +gy3 + hy4

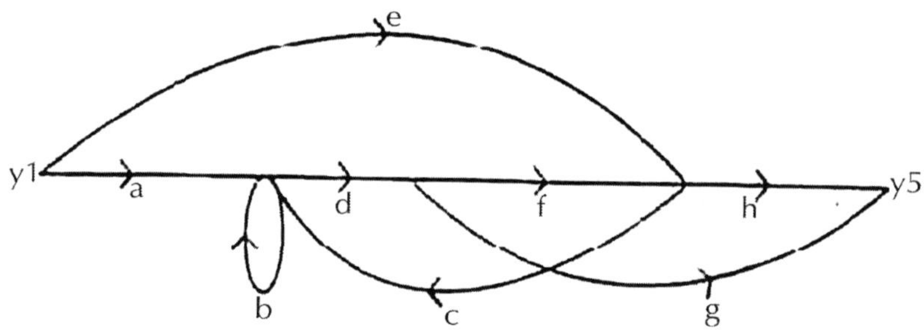

FIGURE 1. Signal flow diagram.

TABLE 1.

Antigen	Intermediate Strength, %	Second Strength, %
Cadidin	39	92
Coccidiodin	19	45
Mixed respiratory vaccine	4	n/a
Mumps	78	n/a
PPD	26	83
SK-SD	55	93
Staphage lystate	71	n/a
Trichophytin	28	n/a

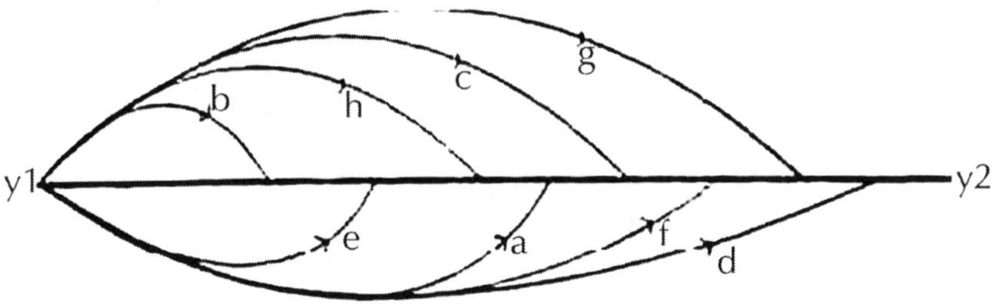

FIGURE 2. Signal flow diagram.

TABLE 2

	Antigen	Rank (%)
a	Candidin	39
b	Mixed respiratory vaccine	19
c	Mumps	41
d	PPD	78
e	SK-SD	26
f	Staphage lysate	55
g	Trichopytin	71
h		28

It is necessary to define the terms as represented by the signal flow diagram in Figure 1. The"source" is a node having only outgoing branches, as y1 in Figure 1. The "sink" is a node having only incoming branches, as y5 in Figure 1. The "path" is a group of connected branches having the same sense of direction. These are he, adfh, and b in Figure 1. The "forward paths" are paths which originate from a source and terminate at a sink along which no node is encountered more than once, as are eh, adg, and adfh in Figure 1. The "path gain" is the product of the coefficient associated with the branches along the path. The "feedback loop" is a path originating from a node and terminating at the same node. In addition, a node cannot be encountered more than once.

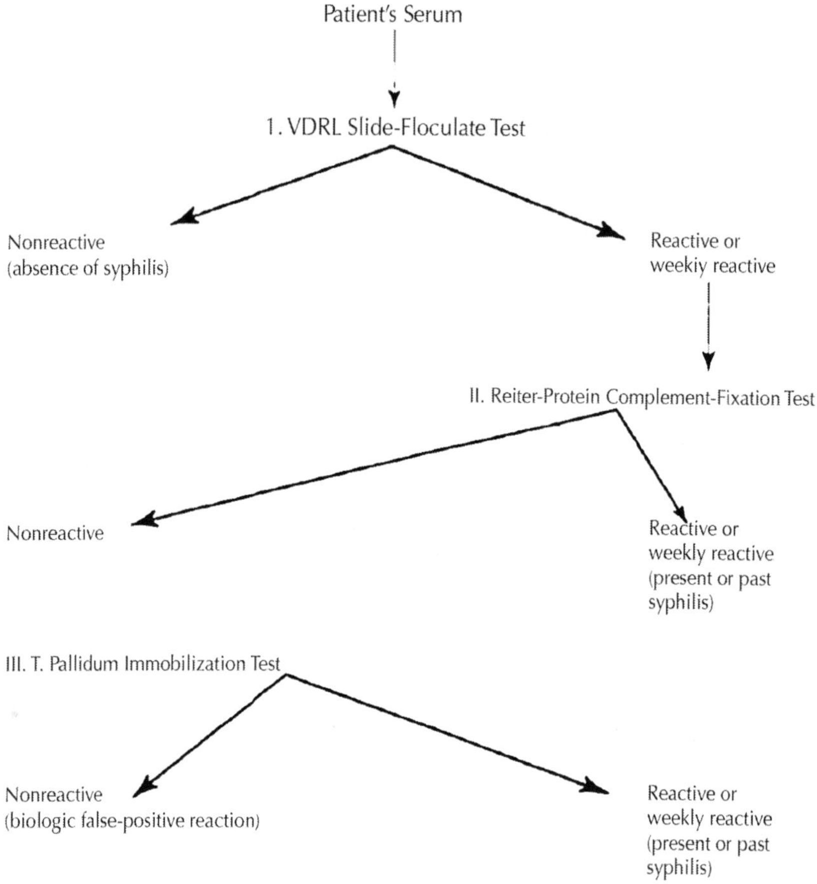

FIGURE 3

They are b and dfc in Figure 1. The "loop gain" is the product of the coefficients associated with the branches fonning a feedback loop.

RESULTS

The perspicuousness or the clarity of the information presented in the signal flow chart, table, and flow chart is the main focus and evaluation criterion of these three systems of infonnation display. Through samples of each system, the strengths of the signal flow diagram are broken down into the following models. Model A is a comparison of a table data and the same data expressed in a signal flow diagram. Model B is a comparison of data in a flow chart and in a signal flow diagram. Model A deals with delayed hypersensitivity skin testing, and model B is the triple-test plan for serology diagnosis in syphilis. Model Cis the zones of inhibition of bacteria by Maguey syrup of the Aztecs, and model D is the molluscicidal activities of selected naphthoquinones from African traditional medicine. Model E is the cyctotoxicity of helenalin and its derivatives from traditional Chinese medicine. In model A the signal flow diagram graphs the rank by percentile of reactions to the six antigens of the hypersensitivity skin test. The sample size was 76 normal adult subjects. The following is the results ofthe six antigens, plus two agents that are also known indicators of possible energy, coccidioidin and mumps, as represented in Table 1. The rank of percentile can be indicated by the signal flow diagram as represented in Figure 2. Table 2 represents the variables (antigens) as they rank in percentile.

From Table 2, and Figure 2 the best antigen for evaluating energy is staphage lysate, at 71%, with SK-SD following in second with 55%. The 78% rating for mumps is greater than staphage lysate, but was not a part of the six antigens scheduled for use. In model B the triple-test plan for serologic diagnosis of syphilis is represented by Figure 3 and has been adapted from a flow chart in ref 9. This flow chart can be represented in a signal flow diagram as represented in Figure 4. The path gain is the line between yl and y5, and y2, y3, and y4 are the test variables. Model C is the use of Maguey sap by the ancient Aztecs for skin infections, utilizing the natural antibiotic and fungistatic activity of saponins. Table 3 shows the zones of inhibition of bacteria by Maguey syrup. The data for the undiluted syrup only is shown as a signal flow diagram in Figure 5.

Model D is the molluscicidal activities of selected naph thoquinones. Schistosomiasi affects millions of people in Africa, Asia, and the South American countries and is transmitted by a number of aquatic snails that play host to miracidia, which in turn is hatched from eggs deposited by humans suffering from the disease. A method of killing the snails is termed molluscicide and the use of naphthoquinones for this process has been validated by laboratory research from the root bark from Diospyros usambarensis. Table 4 is a list of molluscicidal activitiesof selected naphthoquinones. The table is then graphed as a signal flow diagram as seen in Figure 6.

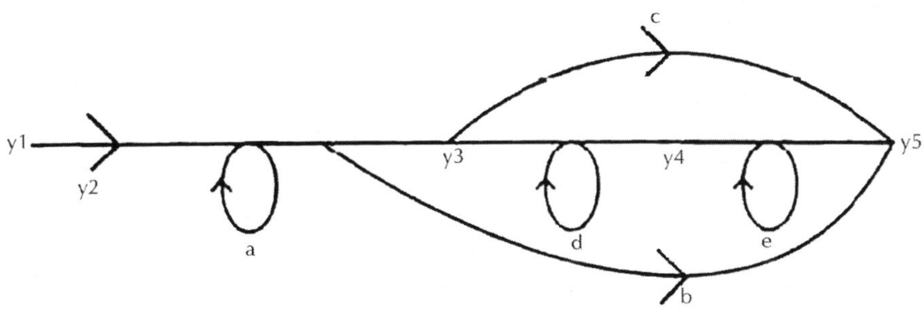

FIGURE 4 Symbol definitions: yl is patient's serum and is the signal source; y2 is the VDRL test; a is the nonreactive response and is a feedback loop; b is reactive and is a path; y3 is the Reiter-Protein Complement Test;c is reactive and is a path; d is nonreactive and is a feedback loop; y4 is T. Palladium Test; e is nonreactive and is a feedback loop; y5 is reactive and is the sink.

TABLE 3 Zones of Inhibition (mm) of Bacteria by Maguey Syrup[a,b]

Bacteria	Undiluted syrup	+ 1.0 mL water	+ 5.0 mL water	+ 0.5mg salt	+ 1.0 mg salt
*Salmonella paratyp*h	50	42	40	46	48
*Pseudomanas aeruginos*a	45	38	38	40	43
Escherichia col	38	27	22	33	38
Shigella sonnel	27	25	25	25	25
Sarcina lutea	23	17	16	21	23
Staphylococcus aureus	20	17	17	33	35

[a]The zone of inhibition is measured as the diameter of the circle.[b]The plus (+) indicates the addition of salt or water to the Maguey syrup. [c]This is a Gram-negative enteric bacteria, rod-shaped. [d]This is a Gram-positive pygenic cocci bacteria.

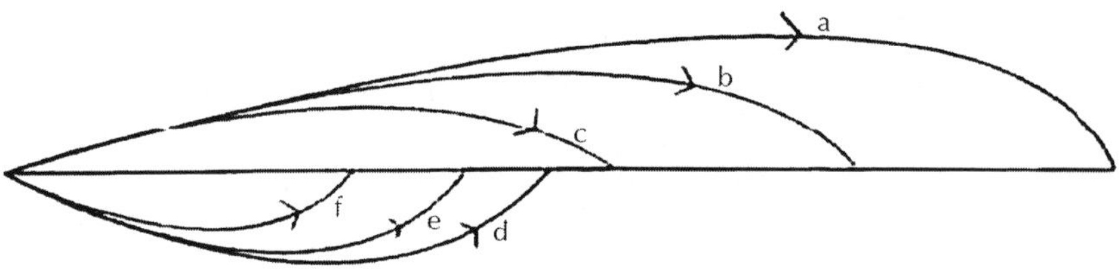

FIGURE 5 Signal flow diagram of zones of inhibition of bacteria by Maguey syrup in undiluted syrup form. Symbol definitions: a is *Salmonella paratyphi*, b is *Pseudomonas aeruginosa*, c is *Escherichia coli*, d is *Shigella sonnei*, e is *Sarcina lutea*,and f is *Staphylococcus aureus*.

TABLE 4

Compound	Mollusidal activity[a] (μg/mL)
7-Methyljuglone	5
Plumbagin	2
Juglone	10

TABLE 4 *(Continued)*

Isojugloine (lawsone)	50
3-Methoxy-7-methyljuglone	50
Vitamin K3 (menadione)	3
Naphthazarin	50

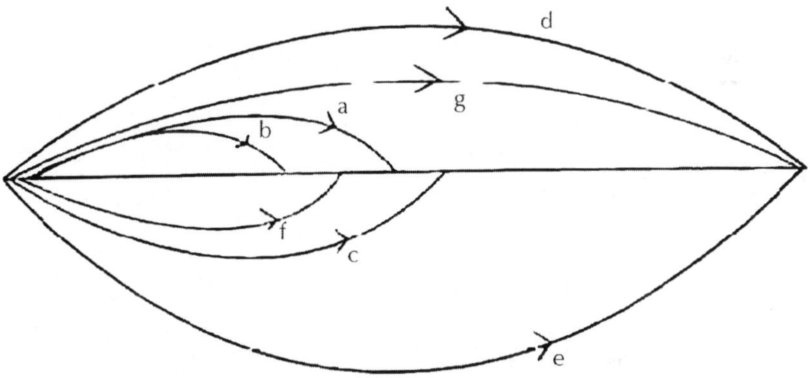

FIGURE 6 Signal flow diagram of molluscicidal activities of selected naphthoquinones. Symbol definitions: a is 7-methyljuglone, b is plumbagin, c is juglone, d is isojugloine (lawsone), e is 3-1~1 ethoxy-7-methyl juglone, f is Vitamin K3 (menadione), and g is naphthazarin.

Model E is the cytotoxicity of helenalin and its derivatives, as hydrogenation of helenalin results in a profound decrease of cytotoxicity and has high antitumor activity against W11-256 ascites carcinosarcoma. Table 6 shows the cytotoxicity ofhelenalin and its derivatives, and Figure 7 is a signal flow diagram of the data in Table 5.

DISCUSSION

In Model A the signal flow diagram shows the rank by percentile of reactions to the six antigens of the hypersensitivity skin test. In this model the contrast is not so much between a table or chart of information but rather how such information would look in a signal flow diagram.

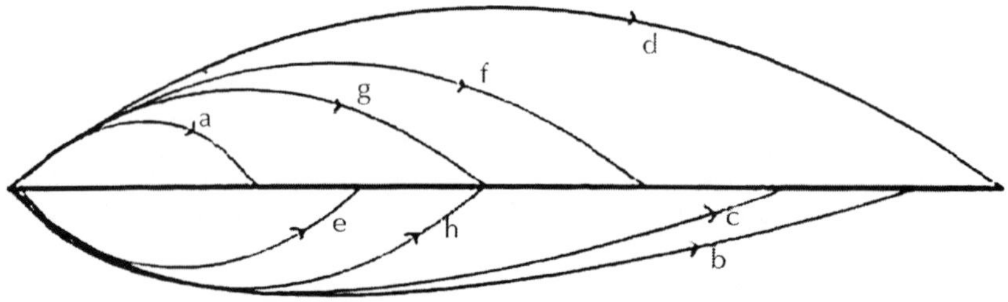

FIGURE 7 Cytotoxicity of helenalin and its derivatives. Symbol definitions: a is helenalin; b is 2,3-dihydrohelenalin; c is 11,13-dihydrohelenal; d is 2,3,11,13-tetrahydrohelenalin; e is 2,3-epoxyhelenalin; f is 1,2-epoxyhelenalin; g is 1,2: 11,13-diepoxyhelenalin; h is 2,3:11,13-diepoxyhelenalin.

TABLE 5

Name	ED50 (HEP-2) (μg/mL)
Helenalin	0.1
2,3-Dihydrohelenalin	3.8
11,13-Dihydrohelenalin	0.81
2,3,11,13-Tetrahydrohelenalin	40.0
2,3-Epoxyhelenalin	0.11
1,2-Epoxyhelenalin	0.53
1,2:11,13-Diepoxyhelenalin	0.50
2,3:11,13-Diepoxyhelenalin	0.50

Upon first inspection of the two systems of representing information, it is apparent that the signal flow diagram is providing a specific type of information more quickly and more clearly than the table of information, mainly that the rank of percentile is more obvious by the use of branches and nodes on the signal flow diagram than is numerically or alphabetically represented in the table. This is a classic case of semantic verses visual information, as the table is a linguistic and numerical device while the signal flow diagram is a visual or graphic device.

In taking into account the space needed to generate the amount of information, the signal flow diagram is superior in that it takes less space, in this case about one quarter of the space of the table and has a clear directional sense, pointing arrows, and a corresponding hierarchy of branches and nodes representing the individual items and their ranking by percent.

In Model B, the triple-test plan for serologic diagnosis of syphilis is first represented by a flow chart and then by a signal flow diagram. The flow chart is a popular method of representing a process hierarchy and is found in most information-oriented disciplines.

The flow chart is a visually clear representation of the information and affords more information than a table or a chart, but as compared to the signal flow diagram, it is again overly complex and time consuming when matched with the simplicity of the signal flow diagram. Notice that the nodes denoting the nonreactive response are clearly represented by loops and that the branches denoting a reactive response are all branched into the reactive sink node of the diagram. This is a more clear representation of the information than can be inferred from the flow chart and takes less space than the flow chart, even with the corresponding table of data.

From the point of view of the information sciences, the signal flow diagram is a superior method of visually displaying mathematical equations in a simple but accurate way. Cybernetics are involved on one level in that the integration of information from raw data to usable information, with complexity, time, and space being constraints on how the information is processed, clearly demonstrates the importance of simple communication systems such as the signal flow diagrams in increasing the efficiency and feedback response (on all levels) of the information sciences.

CONCLUSIONS

1. The signal flow diagram is a simplified graphic representation of mathematical, numerical, and word models, and these models are best expressed by the signal flow diagram.
2. The nodes and branches of the signal flow diagram can symbolize all types of data and information. The branch arrows can represent loops, increases, and decreases in relation to the information being represented, and the nodes denote a hierarchy of the information being represented.
3. Time and space are saved by the use of signal flow diagrams, and this can be important when labor, cost, and efficiency factors are involved.
4. The complexity of the information is made simpler and more visually clear by the use of signal flow diagrams, and this simplicity is inherent in such a graphical method.
5. The signal flow diagram is superior to formal or blocks diagrams, as block diagrams are inherently weak in their mode of simplification and ease of use and are also time and space sensitive.
6. Both equations and raw data are inferior to signal flow diagrams in that equations are long, complex, and time consuming, and raw data is marginal at imparting specific information when compared to the signal flow diagram.
7. The signal flow diagram is superior to tables, charts, and flow charts in that it more readily accepts large quantities of information and represents them in the most accurate and simplistic manner, something that tables, charts, and flow charts achieve with limited success.

8. The information saturation point of signal flow diagrams is higher than other forms of information representation.
9. Overall simplicity of conception, use, and understanding is the main point of interest and support for the signal flow diagram.
10. Accuracy of the signal flow diagram is in the simplicity of its use.

The signal flow diagram is the graphical method of choice for the representation of mathematical, numerical, and word models and is superior to equations, raw data, tables, charts, flow charts, and block diagrams in representing the desired information. Also, the type of information sources can be from nontraditional fields and that the graphic display of such information may have long ranging goals for "updating" the research and testing of such ancient methods of healing and medicine.

CHAPTER 15

CHROMATIC ASPECTS OF THE SIGNAL FLOW DIAGRAM

BRADLEY S. TICE

INTRODUCTION

The use of color in the graphing of nodes and branches of the signal flow diagram can enhance the amount of information conveyed as well as it is perspicuousness. A novel feature of reducing information redundancy is the result of using a chromatic hierarchy in the signal flow diagram.

The signal flow diagram has the conventional form of a circuit diagram, comprising connected branches and nodes, and differs from a circuit diagram m two important ways. The nodes are the points where the signals appear and where they are summated, and that the branches are oriented and their orientation forms a single channel for the travel of the signal (Macmillian, 1964). Mason introduced the signal flow diagram in 1953 as a process for analysis of linear systems as a graphic method of representation of a set of linear algebraic equations (Johnson and Johnson, 1972). When the equations represent a physical system, the graph depicts the flow of signals from one point of the system to another. In 1956, Mason would develop a gain formula that would give a transfer function ofa linear system (Johnson and Johnson, 1972). Coates (1959) flow graph is a generalization of the signal flow graph and depends on the structure of the set of equations (Johnson and Johnson, 1972).

The use of signal flow diagrams are common m fields such as engaging and the information sciences.The main reason for the use of signal flow diagrams over other diagram systems are that they are easy to use and permits a solution practically upon visual inspection (Shinners, 1964) Signal flow diagrams can solve complex linear, multiloop systems in less time than either block diagrams or equations (Macmillian, Higgins, and Naslin, 1964).

A signal flow graph is a topological representation of a set of linear equations as represented by the following equation:

Equation 1 y = a x , 1 = 1,..., n

Branches and nodes are used to represent a set of equations in a signal flow graph. Each node represents a variable in the system, like node "i" represents variable "y" in Equation 1. Branches represent the different variables such as branch "ij" relates variable "yi" to "yj" where the branch originates at node "i" and terminates at node "j" in Equation 1 (Shinners, 1964).

The following set of linear equations are represented in the signal flow graph in Figure 1. (Shinners, 1964).

y2 = ay, +by2 +cy4
y3 = dy2
y4 = ey1 +fy3
y5 = +gy3 + hy4

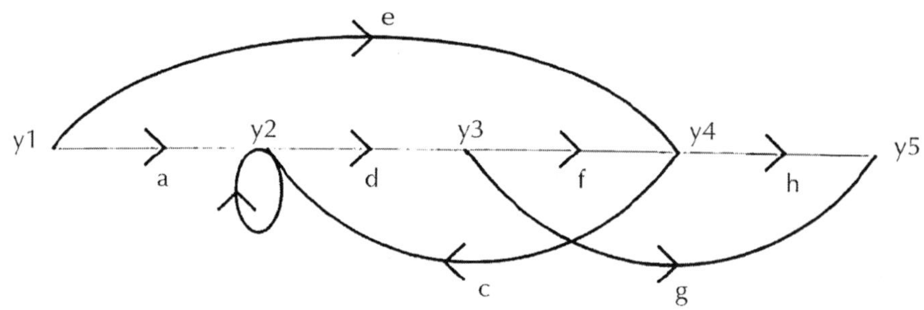

FIGURE 1

It is necessary now to define the terms as represented by the signal flow diagram in Figure 1 (Shinners, 1964).

1. The sourceis a node having only outgoing branches, as "y1" in Figure 1.
2. The sinkis a node having only incoming branches, as "y5" in Figure1.
3. The pathis a group of connected branches having the same sensedirection. These are "he", "adfh", and "b" m Figure 1.
4. The forward pathsare paths which originate from a source and terminate at a sink along which no node is encountered more than once, as are "eh", "adg", and "adfh" in Figure 1.
5. The path gainis the product of the coefficient associated with the branches along the path.
6. The feedback loopis a path originating from a node and terminating at the same node. In addition, a node cannot be encountered more than once. They are "b" and "dfc" in Figure 1.
7. The loop gain is the product of the coefficients associated with the branches forming a feedback loop.

MATERIALS AND METHODS

The materials for this study are as follows:

1. Monochrome signal flow diagram.
2. Signal flow diagram with chromatic branches.
3. Signal flow diagram with chromatic nodes.
4. Signal flow diagram with chromatic branches and chromaticnodes.
5. Signal flow diagram with chromatic branches and no symbols.

The method employed for evaluation is the qualitative evaluation of the data presented and how quickly and accurately this data is analyzed and absorbed by the observer. The analysis of the "perspicuousness" or the clarity of the information presented is the main focus and evaluator ofthese systems of information display.

RESULTS

The graph in Figure 2 is a monochrome signal flow diagram with the variables designated as 'y' and numbered according to placement of each linear equation.

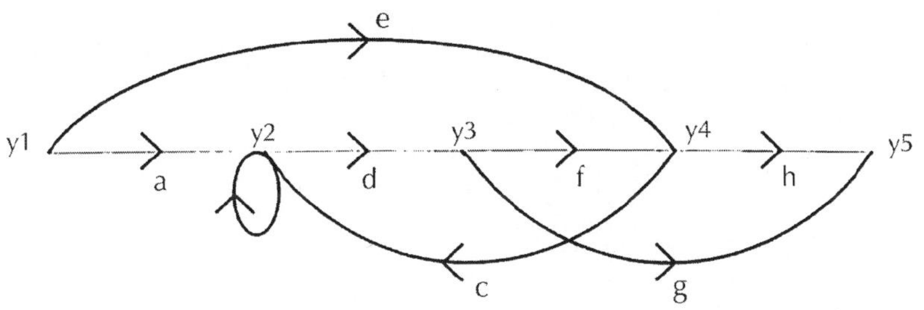

FIGURE 2

The graph in Figure 3 is a signal flow diagram with chromatic branches. The hierarchy of color matches the variables, designated as 'y', and numbered according to placement of each linear equation.

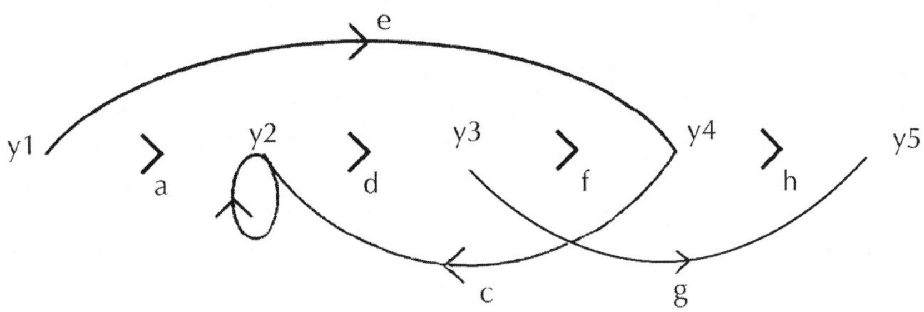

FIGURE 3

The graph in Figure 4 is a signal flow diagram with chromatic nodes. Specific color of node represents both origin and termination point for single arrow graphs and origin and termination point for a common node that is shared by both an arrow and a loop graph.

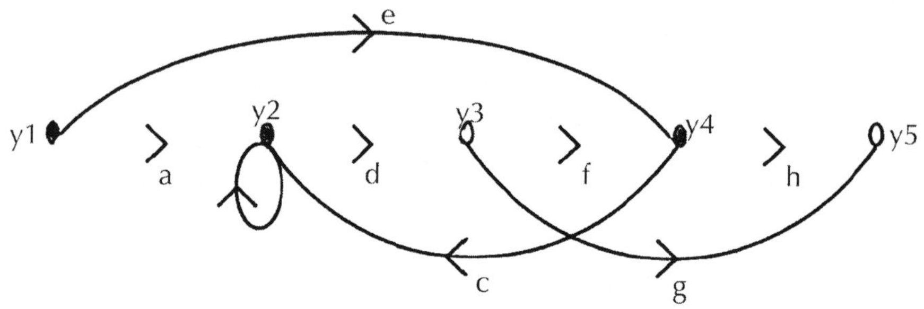

FIGURE 4

The graph in Figure 5 is a signal flow diagram with chromatic branches and chromatic nodes. Chromatic hierarchy of specific colors paralleling branches and nodes as well as common nodes that function for both arrows and loops.

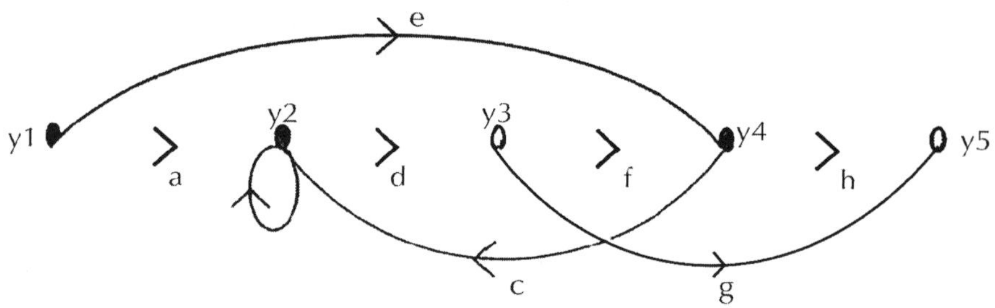

FIGURE 5

The graph in Figure 6 is a signal flow diagram with chromatic branches and no symbols. Because of the specific color of each branch represents a specific equation, the need for marking such equations with corresponding symbols becomesa redundant feature as the specific nature of the equation is already represented by a specific color, removing the need for such symbols. This not only reduces redundancy of the information imparted, but keeps the signal flow diagram a clean and simple graphing device.

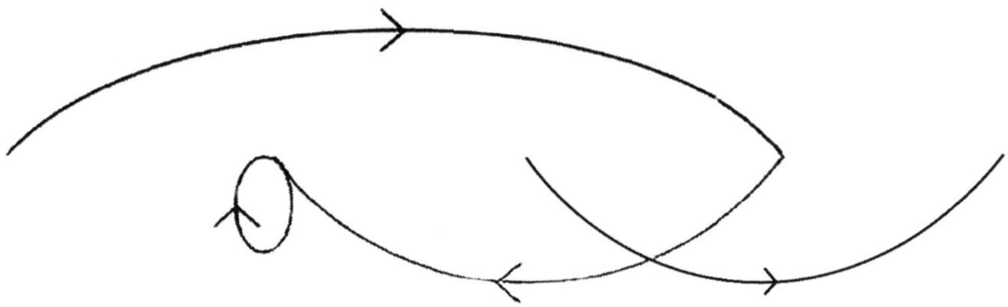

FIGURE 6

DISCUSSION

The following can be said of the strengths of the use of chromatic branches and nodes in the signal flow diagram.

1. The use of specific colored graphs for each branch and node in a signal flow diagram increases the "perspicuousness", or clarity, of that signal flow diagram.
2. The hierarchy of each colored branch and node in the signal flow diagram parallels the hierarchy inherent in that signal flow diagram.
3. Because a natural hierarchy exists that parallels color with each specific linear equation, the denoting of such equations by symbols is a redundant feature and can be discarded in favor of the chromatic branch and node in the signal flow diagram.
4. Because the signal flow diagram can represent many types of data and information, the specific use of color as a symbol to represent a system can be developed. An example would be the use of the color red to symbolize a problem that would occur if a feedback loop develops, hence the graphing of a loop in red to represent such a problem affords moreinformation than justrepresenting an equation as it represents a system. This expands the use of the signal flow diagram from just a representation of a senes of equations to a qualitative tool for analysis of a system.
5. The use of color in branches and nodes in the signal flow diagram can enhance the type and amount of information conveyed as opposed to the standard monochrome models of the signal flow diagram.

CHAPTER 16

JUNCTION GRAPHS

BRADLEY S. TICE

INTRODUCTION

Junction graphs are seamless block diagrams that function as diagrams for process or system pathways. A development from traditional block diagrams to impart more information to graphing systems.

Junction graphs are a development from the “embedded symbol notation linear block diagram” that had as a point of divergence from block diagrams the absence of connecting lines between blocks or cells in favor of “flow” arrows and space directed numeration symbols that served as indicators for the direction of the process pathway of a system (Tice,2008).

Like embedded symbol notation diagrams, junction graphs function without the need of connective lines between blocks to give process direction or “connectedness” usually found in traditional diagrams. While utilizing numeric symbols that are situated relative to the input/output aspect of that particular function within the block or cell, directional arrows are used for process pathway direction within each block (Figure 1).

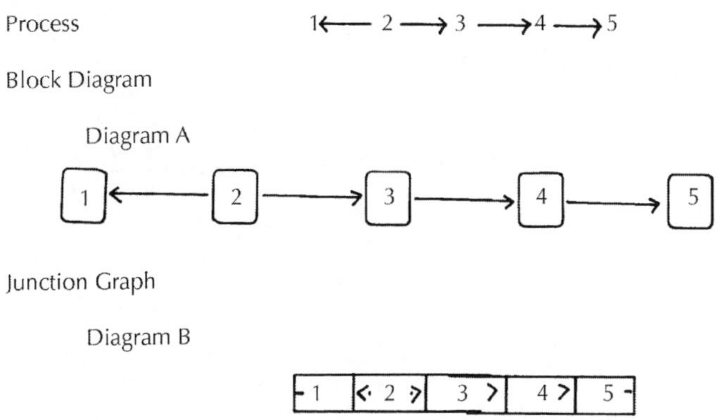

FIGURE 1

KEY TO DIAGRAM B

- < Process flows to the left.
- > Process flows to the right.
- . Point of origin for process flow.
- — Point of process flow termination.

Some observations in comparing diagram A to diagram B.

1. No connecting lines in junction graph between blocks.
2. Space saved in using junction diagram, that is takes up less space than traditional block diagram.
3. Imparts same amount of signal information as traditional block diagram.
4. Removes redundancy of non-connected cells and lines in that junction edges form the "connectedness" of individual cell s and the totality of the system.
5. Visual elements of symbols ; <, >, and —, form a more systematic representation of information codes than traditional block diagram lines between block diagram cells.
6. A cognitive "whole" of the process or system is achieved more readily and perspicuously than traditional block diagrams.

Numeric symbols inside each block or cell, embedded symbol notations, are used to define specific function pathways beyond directional arrow pathways (Figure 2).

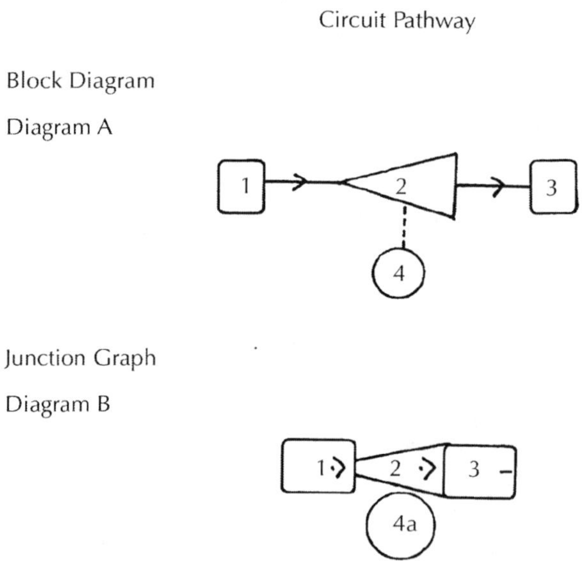

FIGURE 2 Circuit pathway.

KEY TO DIAGRAM B

The sub-notation in cell 4 [4a] is a variable process pathway, that is it becomes a part of the information signal pathway when needed, otherwise is a bypassed function (Tice).

SUMMARY

The junction graph offers the following features of use:

1. Efficiency of graphing space used by removal of redundant connection lines from block diagram and a "seamless" process pathway is invoked by a modular integration of all blocks or cells into a common system.
2. System notations are clear and defined as pathway information has increased by the use of directional arrows and the origin and termination symbols.
3. The use of numeric symbols adds to the "feature" functions of each block and removes the need for descriptive wording around blocks that can be saved for a common key index.
4. Junction graphs have a perceptually pronounced "holistic" feature of describing the totality of the process or system upon visual inspection.

COMMENTARY

Junction graphs fill the need beyond traditional block diagrams in that they are space saving, efficient and impart more information than other graphing systems. As information becomes more complex, the need for visually clear, conceptually accurate graphing techniques will place more demands on traditional modes of graphing. The junction graph was developed with these increases in mind and represent a development toward the future of information systems.

CHAPTER 17

EMBEDDED SYMBOL NOTATION DIAGRAMS AND EMBEDDED SYMBOL NOTATION MATRIX DIAGRAMS

BRADLEY S. TICE

INTRODUCTION

The evaluation of a new system of drafting network circuit systems and process graphs in a manner that is both more efficient and effective than current techniques allow. By using this as a measure of "how" and "what" we perceive as being necessary in a visual graph to represent a process or a system, the very definition of perception comes into focus as an explicit process.

The evaluation of a new system of drafting circuit systems and process graphs in a manner that is both more efficient and effective than current techniques allow. By using this as a measure of "how" and "what" we perceive as being necessary in a visual graph to represent a process or a system, the very definition of perception comes into focus as an explicit process.

The use of embedded symbol notation diagrams and matrixes will be used as a marker for both "what" we perceive as well as "how" we perceive in context to visual images as stimuli for the human perceptive processes. Space, complexity and overall information signal quality will be the standards from which a cognitive model of human perceptual values becomes a quantitative value for a qualitative process of human measurement.

EMBEDDED SYMBOL NOTATION DIAGRAMS

The embedded symbol notation diagram is a self-contained graphic device that incorporates the essence of input and output lines of contact, that is systems are connected by lines, wires, cables, feedback, or just implied transport of information, without the process of connection being manifest in a physical state, that is drawing lines from one system unit to another system unit to represent a point or points of interaction, that take up a substantial amount of space and increases information redundancy.

Embedded symbol notation diagrams have as a notation for such input and output junctions corresponding numbers or alpha symbols that represent the origin of the line, by the underlining of such a number or alpha symbol, and that the corresponding numbers or alpha symbols in other units of the systems diagram are the input junctions from that original line source. From this a corresponding network of connections is developed that need not be physically drawn from unit to unit, thereby removing a redundant feature that becomes obvious upon inspection of this graphing method.

Note: The embedded symbol notation system is a new graphic system designed to increase the quality of systems design by removing redundant or antiquated design features that limit the growth of this design medium. A formal paper developing this system to a more advanced level will be present at a future date.

EMBEDDED SYMBOL NOTATION LINEAR BLOCK DIAGRAMS

A process of the system, the flow of the system from beginning to end, can be done by the use of the embedded symbol notation linear block diagram. This incorporates the use of each unit of the system in a hierarchy of function that is placed in a linear order that follows the "process" of that system from beginning to end. This is realized by the placing of diagram units, or blocks, into a linear pattern that represents to information flow of that process or system.

ORIGINAL DIAGRAM SCHEME

The following is an original diagram scheme of a C-Quam receiver system taken from Gary M. Miller's *Modern Electronic Communication* (Englewood Cliffs: Prentice Hall Carreer & Technology, 1993).

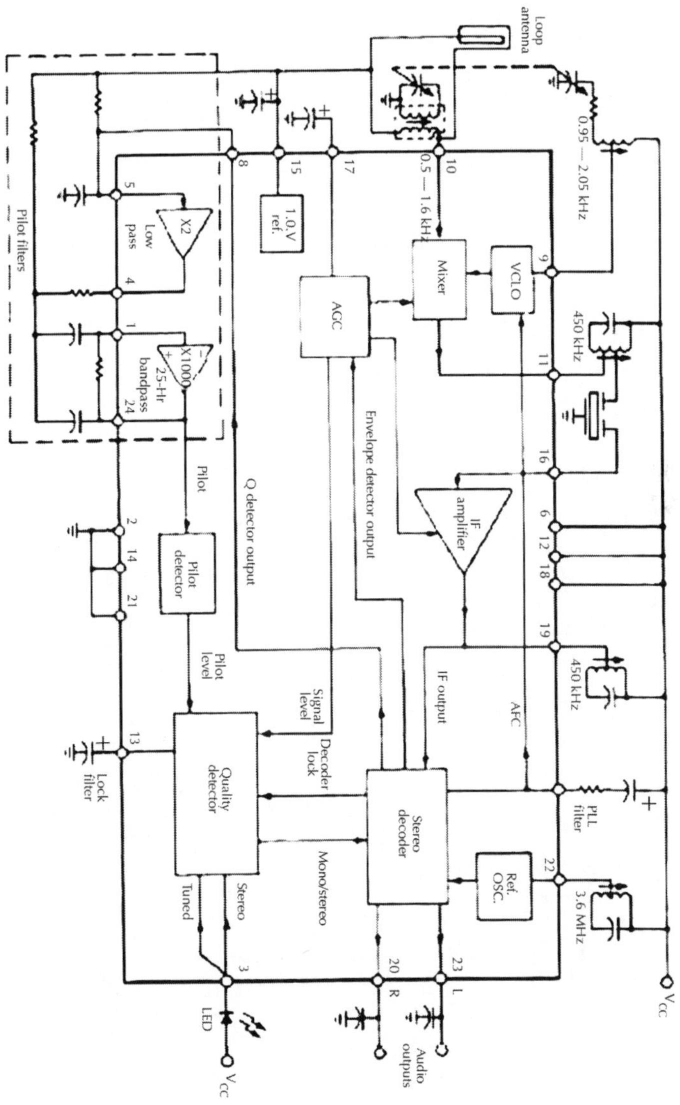

FIGURE 1 C-Quam receiver system.(Courtesy of Motorola Inc.)

EMBEDDED SYMBOL NOTATION DIAGRAM (EXAMPLE)

The following is an example of a embedded symbol notation diagram using the original from Miller's *Modern Electronic Communication* (Miller, 1993). This diagram is standard spaced in that it corresponds to the measure of distance between each units found in the original drawing when either expanded or reduced.

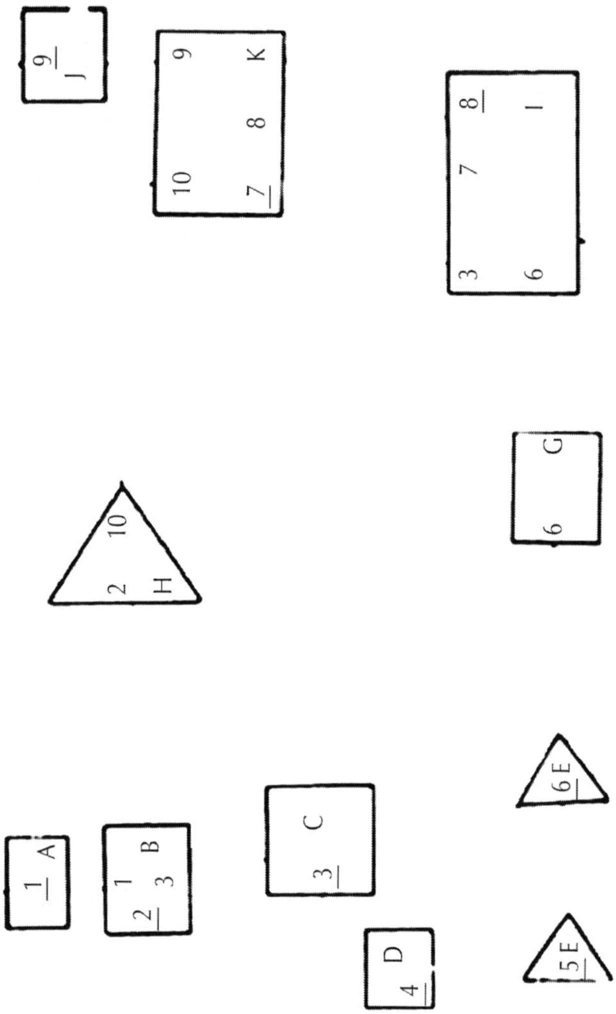

EMBEDDED SYMBOL NOTATION LINEAR BLOCK DIAGRAM (EXAMPLE)

The following is a embedded symbol notation linear block diagram from the original in Miller's *Modern Electronic Communication* (Miller, 1993). Each unit is identified by a corresponding alphabet symbol that denote the hierarchy of its function within the system. The letter [A] corresponds to the beginning or origin of the system followed until the last alphabet letter that represents the end or terminus point of the system.

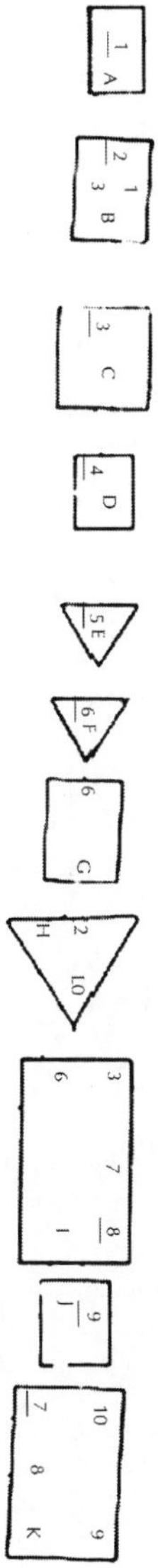

ORIGINAL EXAMPLE NUMBER TWO

The following is another example of a standard graphing system for a generalized circuit connection for FSK and tone detection as taken from Mark E. Oliver's Laboratory Manual for *Modern Electronic Communication* (Englewood Cliffs: Regents/Prentice Hall, 1993).

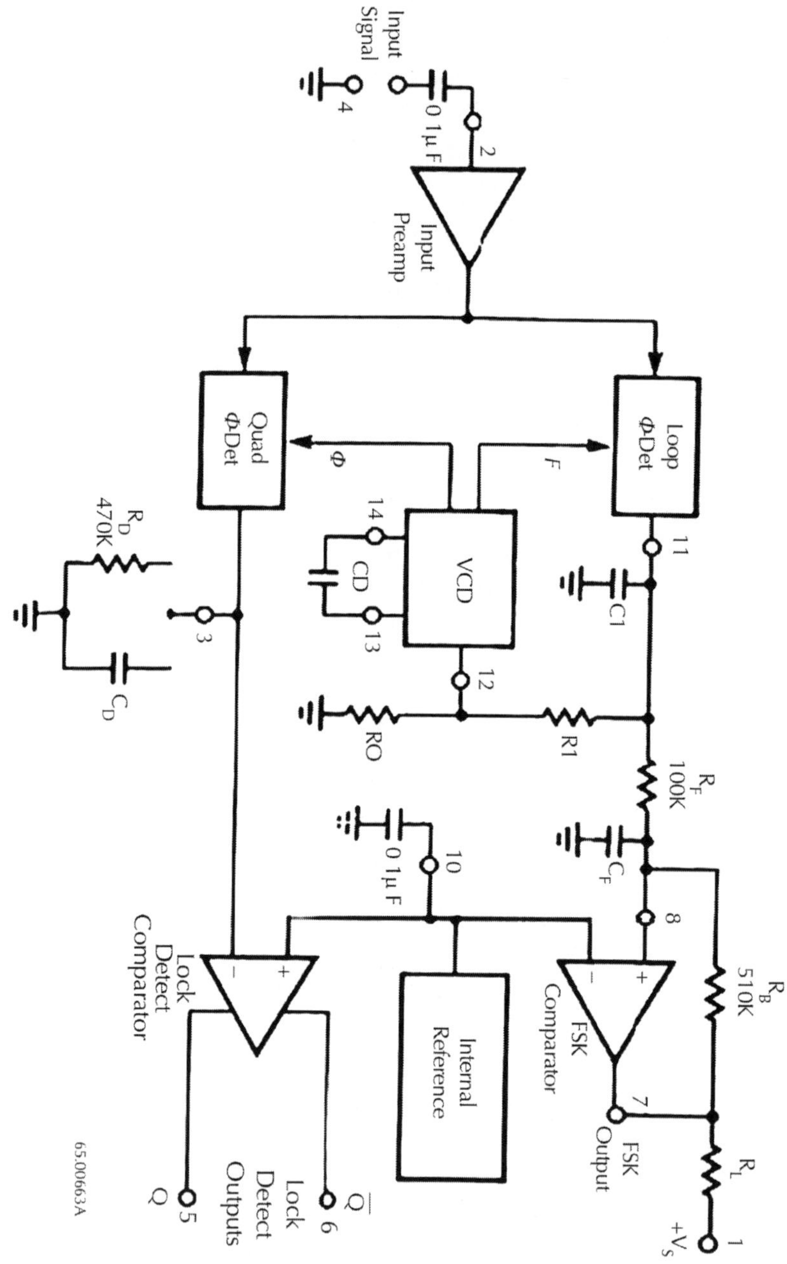

FIGURE 1 Generalized circuit connection for FSK and tone detection.

EMBEDDED SYMBOL NOTATION DIAGRAM EXAMPLE NUMBER TWO

The following is an example of an embedded symbol notation diagram that represents the original diagram taken from Oliver's *Laboratory Manual for Modern Electronic Communication* (Oliver,993).

2. *Note*: A more comprehensive and advanced development of the embedded symbol notation linear block diagram is a "junction graph" that will be presented at a future date.

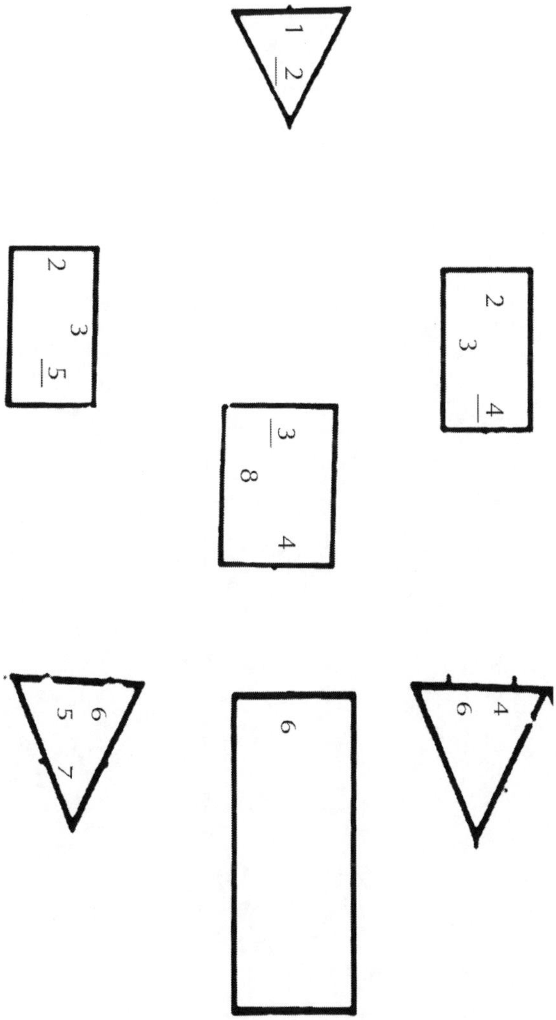

EMBEDDED SYMBOL NOTATION LINEAR BLOCK DIAGRAM EXAMPLE NUMBER TWO

The following is an example of an embedded symbol notation linear block diagram that represents the original diagram taken from Oliver's *Laboratory Manual for Modern Electronic Communication* (Oliver, 1993).

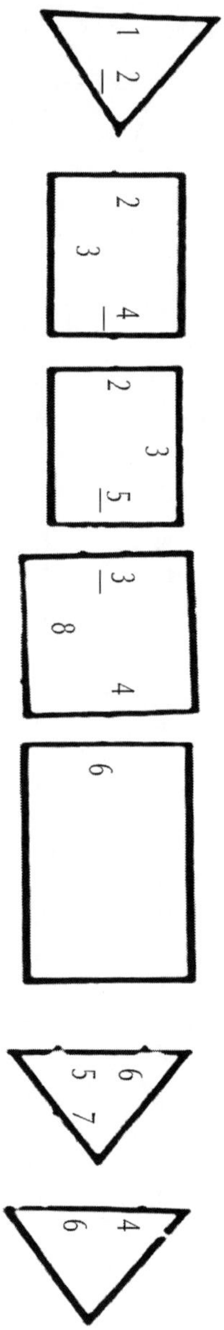

EMBEDDED SYMBOL NOTATION MATRIX

A matrix is a series of linear columns that hold data in a priority manner that gives them a hierarchy and information quality by specific placement on the location of the matrix chart. When they represent a corresponding hierarchy of information from adiagram, the resulting matrix reduces the space needed for such information and makes the graphic elements of the diagram redundant.

ORIGINAL DIAGRAM EXAMPLE NUMBER THREE

The following example of a PLA programming diagram for the 2-bit adder outputs So and cout from Kenneth J. Breeding's *Digital Design Fundamentals* (Englewood Cliffs: Prentice Hall, 1992).

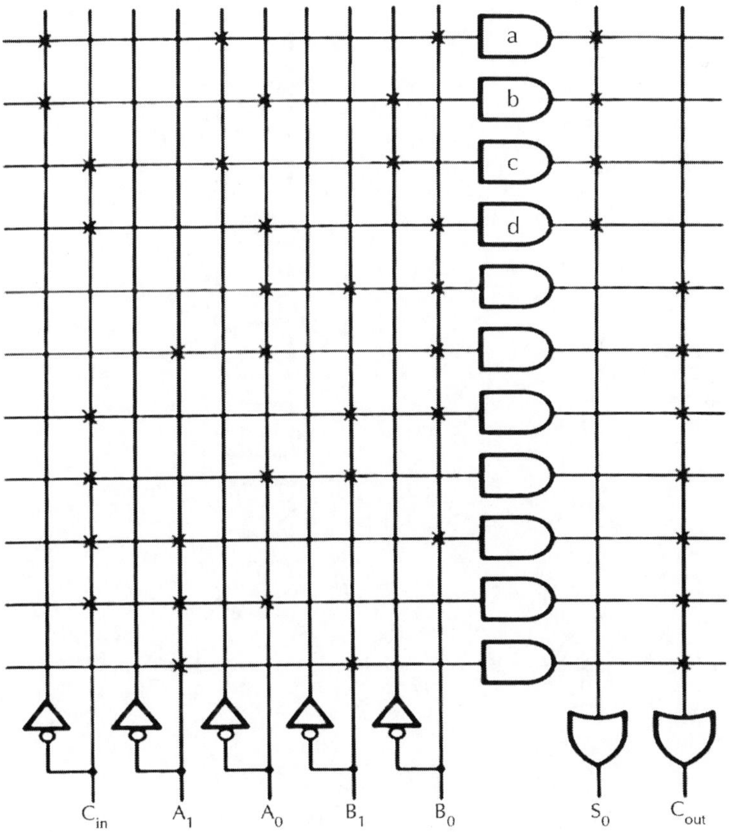

FIGURE 2 The PLA programming diagram for the 2-bit adder outputs S_0 and C_{out}.

EMBEDDED SYMBOL NOTATION MATRIX WITH SCALE

The following is an example of an embedded symbol notation matrix that is done to scale with the original from Breeding's *Digital Design Fundamentals* (Breeding, 1992).

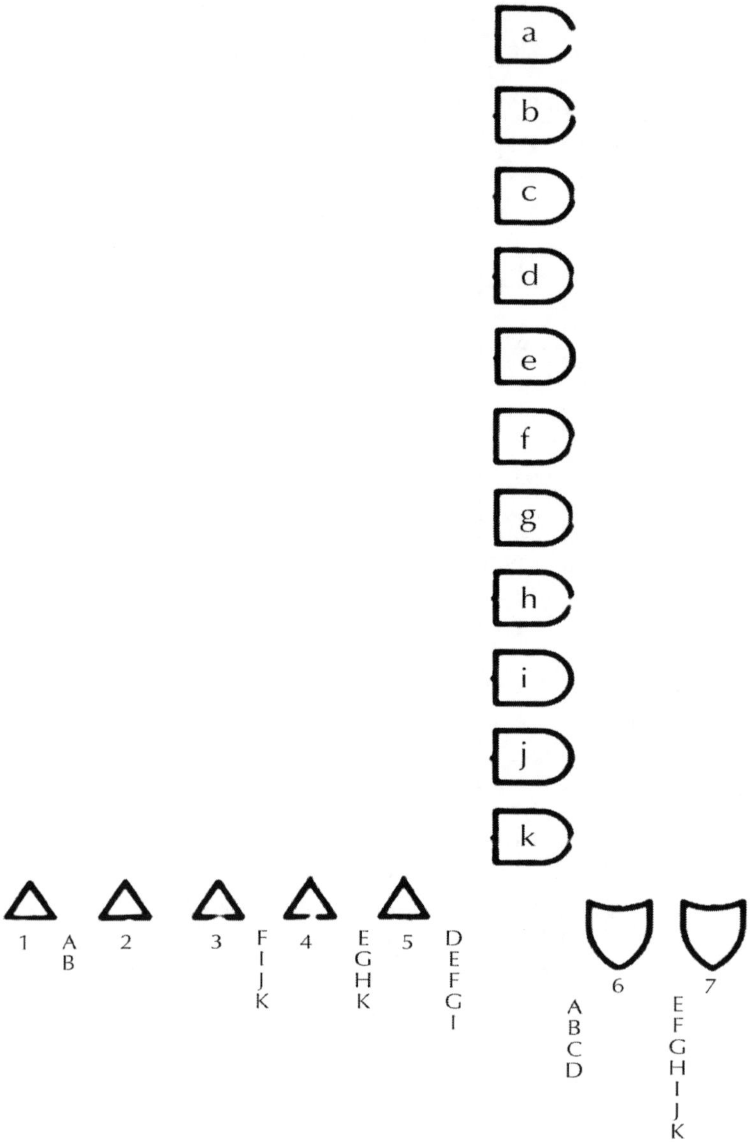

EMBEDDED SYMBOL NOTATION MATRIX

The following is an example of an embedded symbol notation matrix using the original data from Breeding's *Digital Design Fundamentals* (Breeding, 1992).

	Rows					
	A	F	E	D	A	E
	B	I	G	E	B	F
		J	H	F	C	G
Columns		K	K	G	D	H
				I		I
						J
						K

DISCUSSION

In removing the physical lines without removing the "connectedness" of this graph system, a clear line of perceptual processes becomes evident in the resulting evaluation of such a process. If our senses are given some sort of information in which to design a pathway or process route from the input/output of such a system, then the function of such information becomes valid regardless of how established it is in the norms of current usage. What is important is that the correct information about "how" the system works in relation to the visual stimuli imparted from the physical layout of the graph or diagram and how that design functions in respect to the knowledge imparted.

Because graphs are primarily a visual system, the preceptor system used for this process are the eyes, vision, and the brain, that physical manifestation secured in the human skull, to develop a representational aspect of the model, the graph or diagram, of an actual or theoretical process or system. The removal of line of process in a circuit diagram or the alinement of process or system units to act as a functioning pathway does not remove the desired information content and, in fact,increases the validity of such a graphing method by the use ofsuch presentations because they reflect the spatial factors that human's use to "measure" and 'place in context' in the everyday world around them.

The human brain has evolved to "see" beyond lines and simple A to B connectedness to "think" in terms of processes or pathways that are implied rather than being cued. So, the use of such cueing, symbol notations in this case, are a higher order method that should be considered in such professions as engineering where such intellect functions are usually reduced to that of drawing lines, like children in a sandbox.

SUMMARY

Both the embedded symbol notation diagrams and matrixes prove to be valuable aids in designing complex systems without information overload or redundancy that is normally found in graphing systems. To this end it becomes apparent that the very concepts of design, flow of information, and vertical and horizontal space all play an influential role in the construction of such systems as they reflect the innate way in which we perceive the world around us.

The concept of distance as a measurement cue as well as to the very nature of content of information when desiring to present an information signal by the use of an abstract system of language codes number, alpha symbols, and graphic lines, all play a role in how we process such information.

CHAPTER 18

FEEDBACK THEORY: PROPERTIES OF SIGNAL FLOW GRAPHS

SAMUEL J. MASON

INTRODUCTION

The equations characterizing a systems problem may be expressed as a network of directed branches. (The block diagram of a servomechanism is a familiar example.) A study of the topological properties of such graphs leads to techniques which have proven useful, both for the discussion of the general theory of feedback and for the solution of practical analysis problems.

A signal flow graph is a network of directed branches which connect at nodes. Branch jk originates at node j and terminates upon node k,itsdirection is indicated by an arrowhead. A simple flow graph is shown in Figure l(a). This particular graph contains nodes 1, 2, 3, and branches 12, 13, 23, 32, and 33. The flow graph may be interpreted as a signal transmission system in which each node is a tiny repeater station. The station receives signals *via* the incoming branches, combines the information in some manner, and then transmits the result along each outgoing branch. Ifthe resulting signal at node j is called x_j, the flow graph of Figure l(a) implies the existence of a set of explicit relationships

$$
\begin{aligned}
x_1 &= \text{a specified quantity or a parameter} \\
x_2 &= f_2(x_1, x_3) \\
x_3 &= f_3(x_1, x_2, x_3).
\end{aligned}
\tag{1}
$$

The first equation alone would be represented as a single isolated node, whereas the second and third equations, each taken by itself, have the graphs shown in Figure l(b) and Figure l(c). The second equation, for example, states that signal x_2 is directly influenced by signals x_1 and x_3, as indicated by the presence of branches 12 and 32 in the graph.

Mason, S.J. (1955) Feedback Theory – Further Properties of Signal Flow Graphs. Technical Report 303, MIT. Reprint from Proceedings of the I.R.E. Vol. 44, No. 7, July 1956. Reprinted with permission from the Institute of Electrical and Electronics Engineers.

In this report we shall be concerned with flow graph topology,which exposes the structure (Gestalt) of the associated functional relationships, and with the manipulative techniques by which flow graphs may be transformed or reduced, thereby solving or programming the solution of the accompanying equations. Specialization to linear flow graphs yields results which are useful for the discussion of the general theory of feedback in linear systems, as well as for the solution of practical linear analysis problems. Subsequent reports will deal with the formal matrix theory of flow graphs, with sensitivity and stablity considerations, and with more detailed applications to practical problems. The purpose here is to present the fundamentals, together with simple illustrative examples of their use.

THE TOPOLOGY OF FLOW GRAPHS

Topology has to do with the form and structure of a geometrical entity but not with its precise shape or size. The topology of electrical networks, for example, is concerned with the interconnection pattern of the circuit elements but not with the characteristics of the elements themselves. Flow graphs differ from electrical network graphs in that their branches are directed. In accounting for branch directions, we shall need to take an entirely different line of approach from that adopted in electrical network topology.

CLASSIFICATION OF PATHS, BRANCHES, AND NODES

As a signal travels through some portion of a flow graph, traversing a number of successive branches in their indicated directions, it traces out a path. In Figure 2, the sequences 1245, 2324, and 23445 constitute paths, as do many other combinations. In general, there may be many different paths originating at a designated node j and terminating upon node k, or there may be only one, or none. For example, no path from node 4 to node 2 appears in Figure 2. If the nodes of a flow graph are numbered in a chosen order from 1 to n, then we may speak of a forward path as any path along which the sequence of node numbers is increasing, and a backward path as one along which the numbers decrease. An open path is one along which the same node is not encountered more than once. Forward and backward paths are evidently open.

Any path which returns to its starting node is said to be closed. Feedback now enters directly into discussion for the first time with the definition of a feedback loop as any set of branches which forms a closed path. The low graph of Figure 2 has closed paths 232 (or 323) and 44. Multiple encirclements such as 23232 or 444 also constitute closed paths but these are topologically trivial. Notice that some paths, such as12324, are neither open nor closed.

We may now classify the branches of a flow graph as either feedback or cascade branches. A feedback branch is one which appears in a feedback loop. All others are called cascade branches. Returning to Figure 2, we see that 23, 32, and 44 are the only feedback branches present. If each branch in a flow graph is imagined to be a one-way street, then a lost automobilist who obeys the law may drive through Feedback Streetany number of times but he can traverse Cascade Boulevard only once as he wanders about in the graph.

The nodes in a flow graph are evidently susceptible to the same classification as branches,that is, a feedback node is one which enters a feedback loop. Two nodes or branches are said to be coupled if they lie in a common feedback loop. Any node not in a feedback loop is called a cascade node. Two special types of cascade nodes are of interest. These are sources and sinks. A source is a node from which one or more branches radiate but upon which no branches terminate. A sink is just the opposite, a node having incoming branches but no outgoing branches. Figure 2 exhibits feedback nodes 2, 3, 4, a source 1, and a sink 5. It is possible, of course, for a cascade node to be neither a source nor a sink. The intermediate nodes in a simple chain of branches are examples.

CASCADE GRAPHS

A cascade graph is a flow graph containing only cascade branches. It is always possible to number the nodes of a cascade graph in a chosen sequence, called the order of flow, such that no backward paths exist.

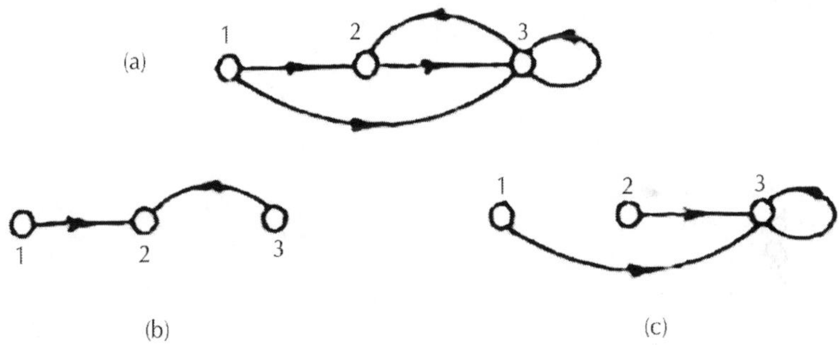

FIGURE 1 Flow graphs.

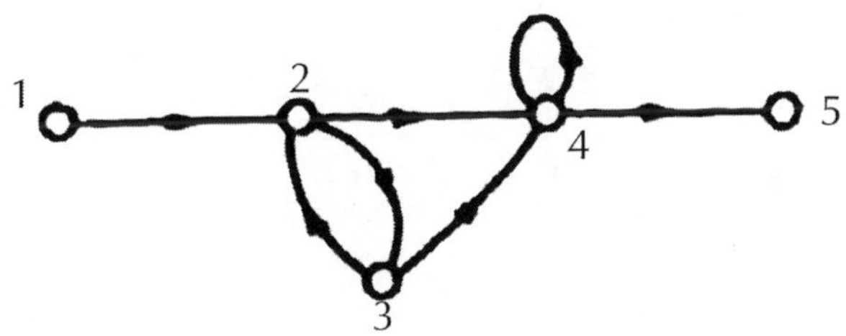

FIGURE 2 A flow graph with three feedback branches and four cascade branches.

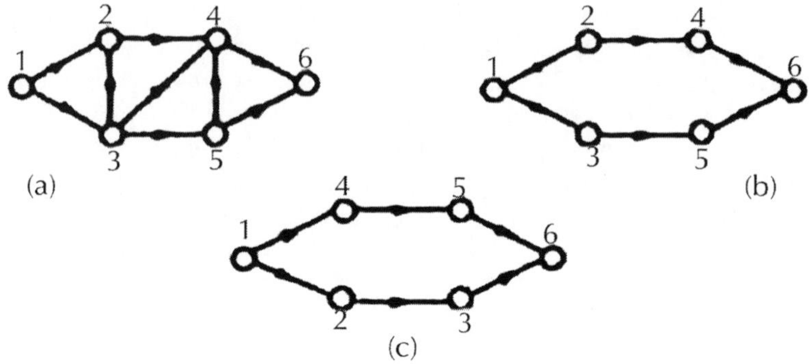

FIGURE 3 Cascade graphs.

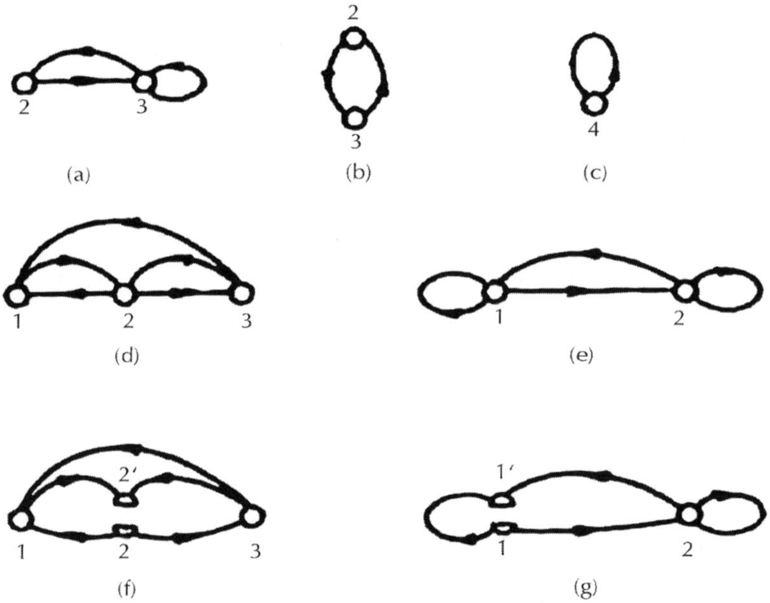

FIGURE 4 Feedback units.

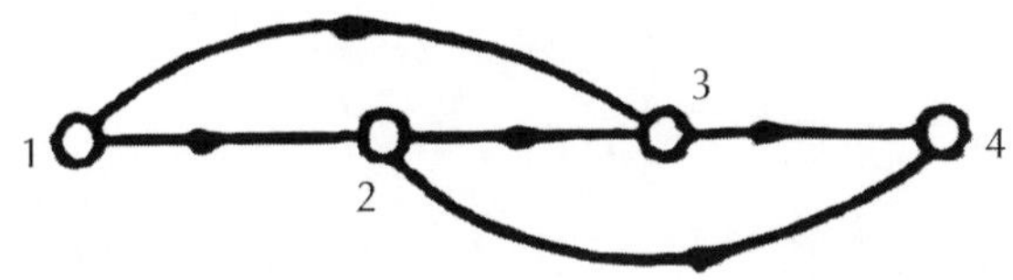

FIGURE 5 A cascade graph.

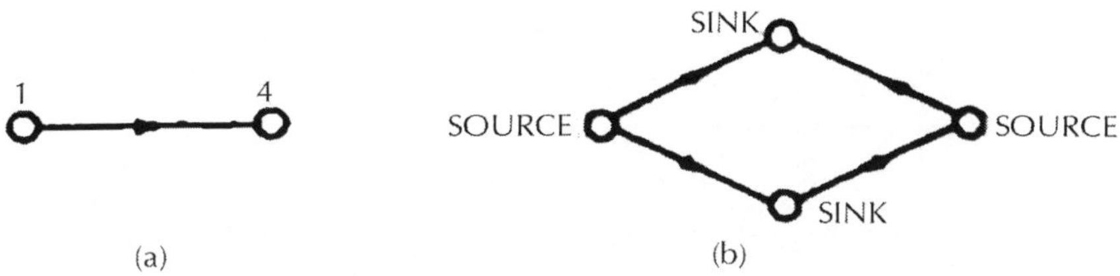

FIGURE 6 Residual forms of a cascade graph.

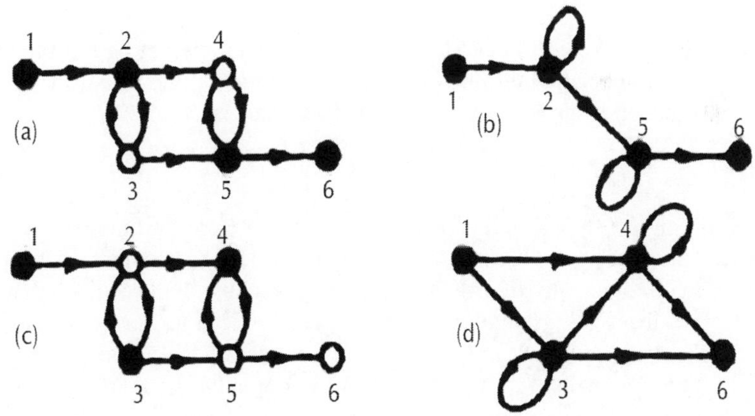

FIGURE 7 Feedback graphs and the index-residues.

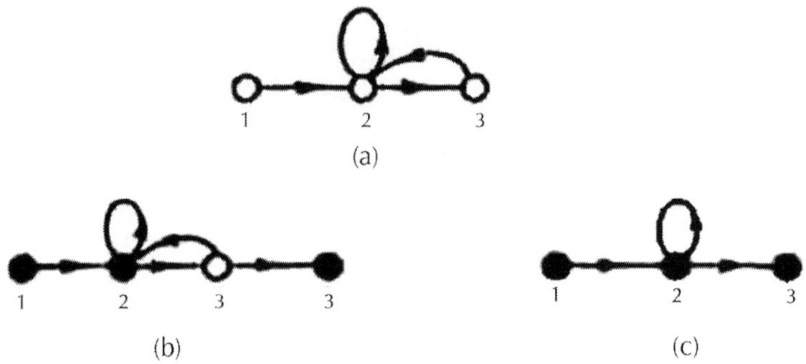

FIGURE 8 Retention of a desired node as a link.

For a proof of this we first observe that a cascade graph must have at least one source node. Let us choose a source, number it one, and then remove it, together with all its radiating branches. This removal leaves a new cascade graph having, itself, at least one source. We again choose a source, number it two, and continue the process until only isolated nodes remain. These remaining nodes are the sinks of the original graph and they are numbered last. It is evident that this procedure establishes an order of flow.

Figure 3 shows two simple cascade graphs whose nodes have been numbered in flow order. The numbering of graph 3(a) is unique, whereas other possibilities exist for graph 3(b); the scheme shown in graph 3(c) offers one example.

FEEDBACK GRAPHS

A feedback graph is a flow graph containing one or more feedback nodes. A feedback unit is defined as a flow graph in which every pair of nodes is coupled. It follows that a feedback unit contains only feedback nodes and branches. If all cascade branches are removed from a feedback graph, the remaining feedback branches form one or more separate feedback units which are said to be imbedded or contained in the original flow graph. The graph of Figure 1, for example, contains the single unit shown in Figure 4(a), whereas the two units shown in Figure 4(b) and (c) are imbedded in the graph of Figure 2.

The units shown in Figure 4(d) and (e) each possess three principal feedback loops. The riumber of loops, however, is not of great moment. A more important characteristic is a number called the index. Preparatory to its definition, let us introduce the operation of node-splitting, which separates a given node into a source and a sink. All branch tails appearing at the given node must, of course, go with the source and all branch noses with the sink. The result of splitting node Z in Figure 4(d) is shown in Figure 4(f). Similarly, Figure 4(g) shows node 1 of Figure 4(e) in split form. We shall retain the original node number for both parts of the split node, indicating the sink by a prime. Splitting effectively interrupts all paths passing through a given node and makes cascade branches of all branches connected to that node.

We can now conveniently define the index of a feedback unit as the minimum number of node-splittings required to interrupt all feedback loops in the unit. For the determination of index, splitting a node is equivalent to removing that node, together with all its connecting branches.

The index of the graph in Figure 4(d) is unity, since all feedback loops pass through node z. Graph 4(e), on the other hand, is of index two.

THE RESIDUE OFA GRAPH

A cascade graph represents a set of equations which may be solved by explicit operations alone. Figure 5, for example, has the associated equation set:

$x_2 = f_2 (x_1)$
$x_3 = f_3 (x_1, x_2)$
$x_4 = f_4 (x_2, x_3)$. (2)

Given the value of the source x_1, we obtain the value of x_4 by direct substitution

$$x_4 = f_4\{f_2(x_1), f_3[x_1, f_2(x_1)]\} = F_4(x_1). \tag{3}$$

In general, there may be s different sources. Once an order of flow is established, a knowledge of the source variables $x_1, x_2, ..., x_s$ fixes the value of x_{s+1}, since no backward paths from later nodes to x_{s+1} can exist. Similarly, with $x_1, x_2, \ldots, x_{s+1}$ known, x_{s+2} is determined explicitly, and so on to the last node x_n. A cascade graph is immediately reducible, therefore, to a residual from in which only sources and sinks appear. The residual form of Figure 5 is the single branch shown in Figure 6(a), which represents Equation 3.Had two sources and two sinks appeared in the original graph, the residual graph would have contained, at most, 4branches, as indicated by Figure 6(b).

Unlike those associated with a cascade graph, the equations of a feedback graph are not soluble by explicit operations. Consider the sample example shown in Figure 1. An attempt to express x_3 as an explicit function of x_1 falls because of the closd chain of dependency between x_2 and x_3. Elimination of x_2 from Equation 1 by substitution yields:

$$x_3 = f_3[x_1, f_2(x_1, x_3), x_3] = F_3(x_1, x_3). \tag{4}$$

Although, a feedback graph cannot be reduced to sources and sinks by explicit means, certain superfluous nodes may be eliminated, leaving a minimum number of essential implicit relationships exposed.

In any contemplated process of graph reduction, the nodes to be retained in the new graph are called residual nodes. It is convenient to define a residual path as one which runs from a residual node to itself or to another residual node, without passing through any residual nodes. The residual graph, or residue, has a branch jk if, and only if,the original graph has one or more residual paths from j to k. This completely defin-esthe residue of any flow graph for a specified set of residual nodes.

We are interested here in a reduction which can be accomplished by explicit operations alone. The definition of index implies the existence of a set of index nodes, equal in number to the index of a graph, whose splitting interrupts all feedback loops in the graph. The set is not necessarily unique. Once a set of index nodes has been chosen, however, all other nodes except sources and sinks may be eliminated by direct substitution, leaving a residual graph in which only sources, sinks, and index nodes appear. We shall call each a graph the index-residue of the original graph.

Figure 7 shows a flow graph (a) and its index-residue (b). Residual nodes are blackened. Branch 25 in (b) accounts for the presence of residual paths 245 and 235 in(a). All paths from 2 to 6 in (a) pass through residual node 5. Hence graph 7(a) has no residual paths from 2 to 6, since a residual path, by definition, may not pass through a residual node. Accordingly, graph 7(b) has no branch 26. Figure 7(c) illustrates an-alternate choice of index nodes and Figure 7(d) shows the resulting index-residue. Choice (a) is apparently advantageous in that it leads to a simpler residue.

A minor dilemma arises in the reduction process if we desire, for some reason, to preserve a node which is neither an index node nor a sink. In Figure 8(a), for example, suppose that an eventual solution for x_3 in terms of x_1 is required. A node corresponding to variable x_3 must be retained in the residual graph. Apparently, no further reduction is possible. The simple device shown in Figure 8(b) may be employed, however, to obtain the residue (c). The trick is to connect node 3 to a sink through a branch representing the equation $x_3 = x_3$. The original node 3 then disappears in the reduction, leaving the desired value of x_3 available at the sink. This trick is simple but topologically nontrivial.

THE CONDENSATION OF A GRAPH

The concept of an order of flow may be applied, in modified form, to a feedback graph as well as to a cascade graph. Consider the feedback graph in Figure 9(a), which contains two feedback units. If each imbedded feedback unit is encircled and treated as a single supernode, then the graph condenses to the form shown in Figure 9(b), where-supernodes are indicated by squares. Since, the condensation is a cascade structure, an order of flow prevails. Within each supernode the order is arbitrary, but we shall agree to number the internal nodes consecutively.

The index-residue of a flow graph shows the minimum number of essential variables which cannot be eliminated from the associated equations by explicit operations. The condensation of the residue programs the solution for these variables. In Figure 9(b), for example, the condensation directs us to specify the value of x_1, to solve a pair of simultaneous equations for x_2 and x_3, to solve a single equation for x_4, and to compute x_5 explicitly. The complexity of the solution, without regard for the specific character of the mathematical operations involved, is indicated by the number of feedback units and the index of each, since the index of a feedback unit is the minimum number of simultaneous equations determining the variables in that unit.

Carrying the condensation one step further, we may indicate the basic structural character of a given flow graph by a simple listing of its nodes in the order of condensed signal flow, with residual nodes underlined and feedback units overlined. The sequence,

$$\underline{1}\quad \underline{2}\quad \overline{3\ \underline{4}\ \underline{5}\ 6}\quad 7\quad \overline{\underline{8}\ 9\ 10}\quad 11\quad \underline{12}$$

For example, states that nodes 1 and 2 are sources, 7 and 11 are cascade nodes, and 12 is a sink. Also nodes 3, 4, 5, and 6 lie in a feedback unit of index two, have index nodes 4 and 5. Finally, nodes 8, 9, and 10 comprise a later feedback unit of index one, 8 being the index node.

THE INVERSION OF A PATH

A single constraint or relationship among a number of variables appears topologically as a cascade graph containing one sink and one or more sources. Figure 10 (a) is an elementary example. At least in principle, nothing prevents us from solving theequation in Figure 10(a) for one of the independent variables, say x_1, to obtain the form shown in Figure 10(b).In terms of the flow graph, we say that branch 14 has beeninverted.

By definition, the inversion of a branch is accomplished by interchanging the nose and tail of that branch and, in moving the nose, carrying along all other branch noses which touch it. The tails of other branches are left undisturbed. The inversion of a path is effected by inverting each of its branches.

Figure 11 shows (a) a flow graph, (b) the inversion of an open path 1234, and (c) the inversion of a feedback loop 343. To obtain (c) from (a), for example, we first change the directions of branches 34 and 43. Then we grasp branch p by its nose and move the nose to node 4, leaving the tail where it is. Finally, the nose of branch q is shifted to node 3. Branches 12 and 32 are unchanged since they have properly minded their own business and kept their noses out of the path inversion. Topologically, the two parallel branches running from 4 to 3 are redundant. One such branch is sufficient to indicatethe dependency of $x_3$upon x_4.

The inversion of an open path is significant only if that path starts from a source. Otherwise, two expressions are obtained for the same variable and 2 nodes with the same number would be needed in the graph. In addition, inversion is not applicable to a feedback loop which intersects itself. The reason is that two of the path branches-would terminate upon a commonnode. Hence, the inversion of one would move the other,there by destroying the path to be inverted. Such paths as 234 and 23432 in Figure 11(a), therefore, are not candidates for inversion.

The process of inversion, as might be expected, influences the topological properties of a flow graph. Of greatest interest here is the effect upon the index. Graphs (a), (b), and (c) of Figure 11 have indices of two, zero, and one, respectively. In general, paths parallel to a given path contribute to the formation of feedback loops when the given path is inverted, and conversely. Hence, if we wish to accomplish a reduction of index we should choose for inversion a forward path having many attached backward paths butfew parallel forward paths.

THE ALGEBRA OF LINEAR FLOW GRAPHS

A linear flow graph is one whose associated equations are linear. The basic linear flow graph is shown in Figure 12. Quantities a and b are called the branch transmissions, or branch gains. Thinking of the flow graph as a signal transmission system,we may associate each branch with a unilateral amplifier or link. In traversing any branch the signal is multiplied, of course, by the gain of that branch.

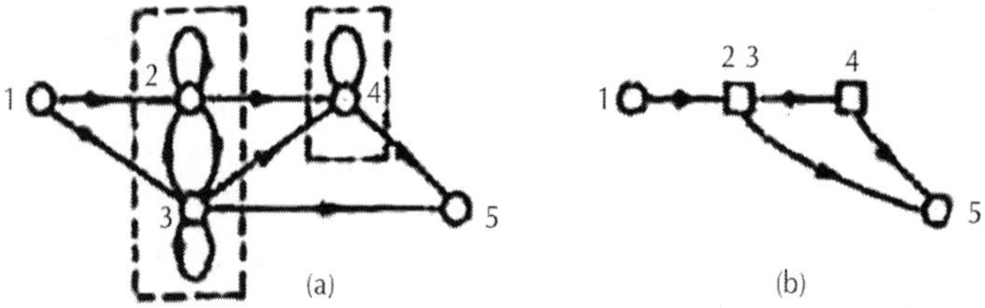

FIGURE 9 The condensation of a flow graph.

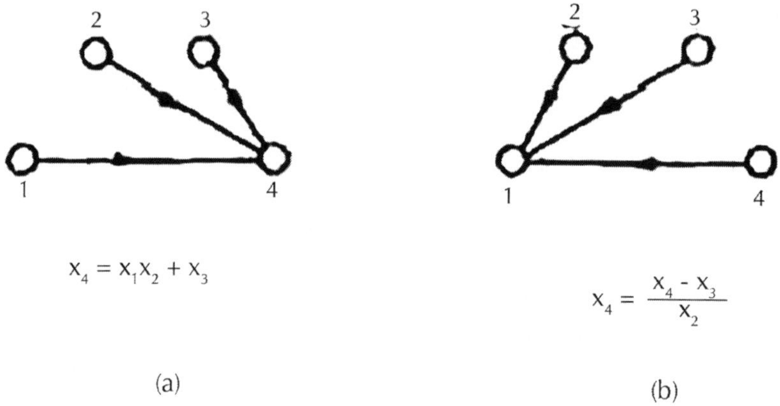

FIGURE 10 Inversion of a branch.

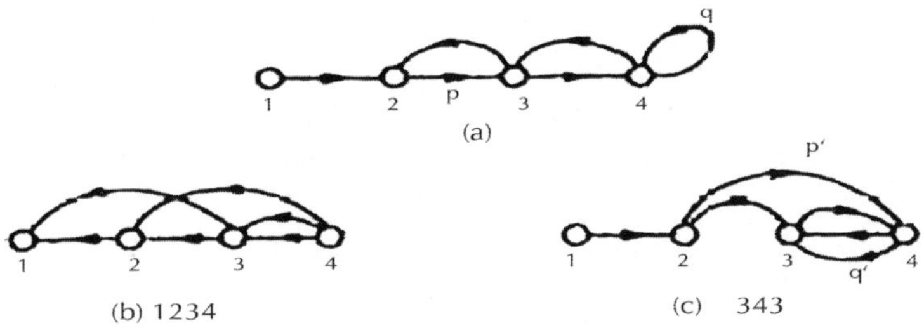

FIGURE 11 Path inversion.

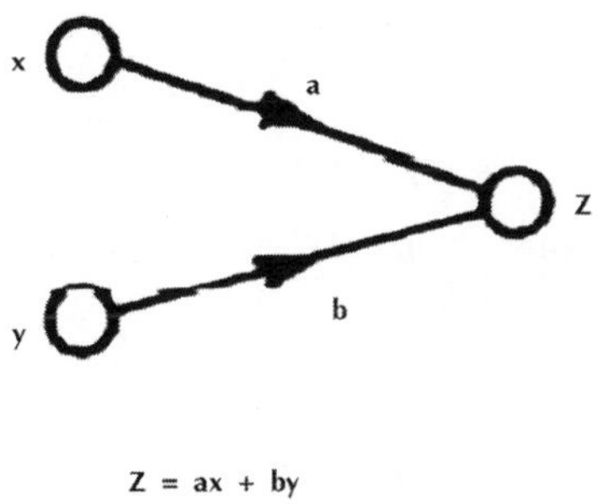

FIGURE 12 The basic linear flow graph.

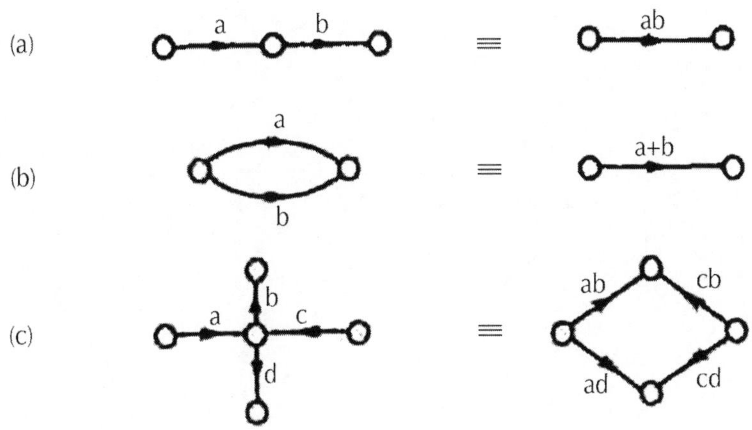

FIGURE 13 Elementry transformations.

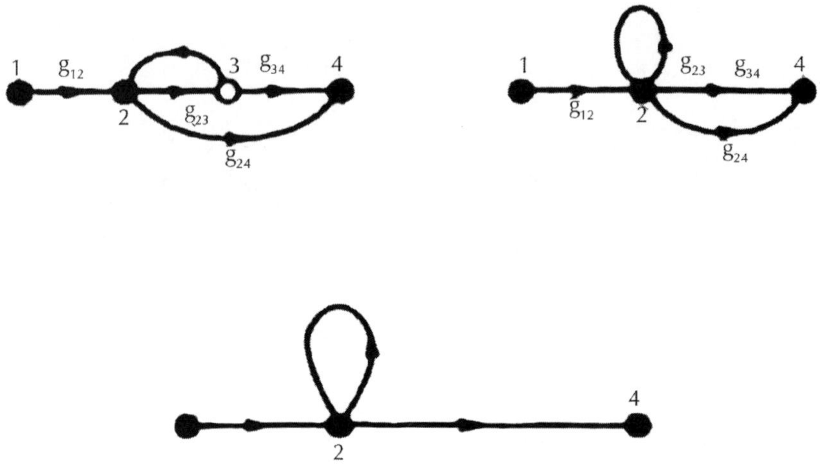

FIGURE 14 Reduction to an index-residue by elementary transformations.

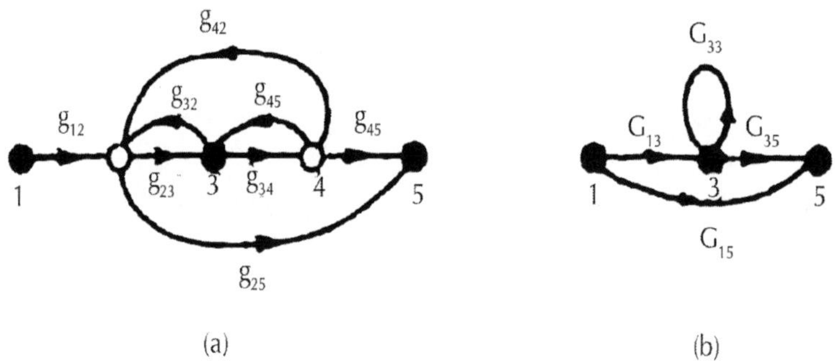

FIGURE 15 Reduction to an index-residue by inspection.

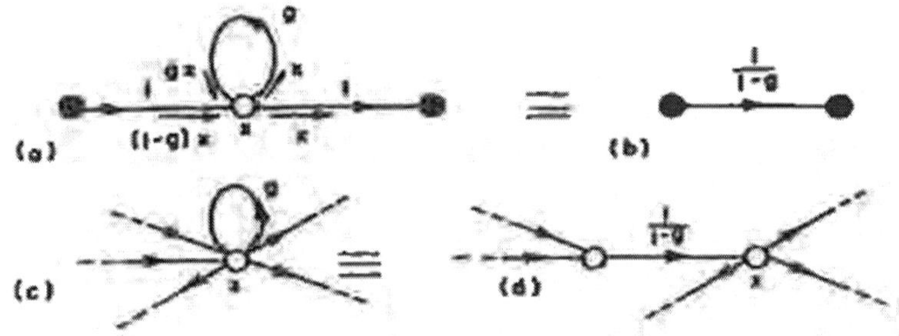

FIGURE 16 Replacement of a self-loop by a branch.

Each node acts as an adder and ideal repeater which sums the incoming signals algebraically and then transmits the resulting signal along each outgoing branch.

ELEMENTARY TRANSFORMATIONS

Figure 13 illustrates certain elementary transformations or equivalences. The cascade transformation (a) eliminates a node, as does the star-to-mesh transformation (c), of which (a) is actually a special case. The parallel or multipath transformation (b) reduces the number of branches. These basic equivalences permit reduction to an index-residue and give us, as a result of the process, the values of branch gains appearing in the residual graph. Figure 14 offers an illustration. The residual nodes are the source 1, the sink 4, and the index node 2. Node 3 could be chosen instead of node 2 but this would lead to a more complicated residue. The star-to-mesh equivalence eliminates node 3 in graph 14(a) to give graph 14(b). The multipath transformation then yields the residue (c).

For more complicated structures the repeated use of many successive elementary transformations is tedious. Fortunately, it is possible under certain conditions to recognize the branch gains of a residue by direct inspection of the original diagram. In order to provide a sound basis for the more direct process, we shall define a path gain as the product of the branch gains along that path. In addition, the residual gain G_{jk} is defined as the algebraic sum of the gains of all different residual paths from j to k. As defined previously, a residual path must not pass through any of the residual nodes which are to be retained in the new graph. It follows that each branch gain of the residue is equal to the corresponding residual gain G_{jk} of the original graph. Moreover, if the residual graph is an index-residue, then each G_{jk} is the gain of a cascade structure and contains only sums of products of the original branch gains. For index-residues, therefore, the gains G_{jk} are relatively easy to evaluate by inspection.

The feedback graph of Figure 15(a), for example, has an index-residue (b) containing four branches. By inspection of the original graph, the residual gains are found to be,

$$G_{13} = g_{12}g_{23}$$

$$G_{15} = g_{12}g_{25}$$

$$G_{33} = g_{32}g_{23} + g_{34}g_{42}g_{23} + g_{34}g_{43}$$

$$G_{35} = g_{34}g_{45} + g_{32}g_{25} + g_{34}g_{42}g_{25} \tag{5}$$

Notice that there are three different residual paths from node 3 to itself and also from 3 to 5. We must be very careful to account for all of them. There is only one residual path from 1 to 5, however, and this is 125. Path 12345, which we might be tempted to include in G_{15}, is not residual, since it passes through node 3.

THE EFFECT OF A SELF-LOOP

When a feedback graph is simplified to a residue containing only sources, sinks, and index nodes, one or more self-loops appear. The effect of a self-loop at any node upon the signal passing through that node may be studied in terms of Figure 16(a).The signal existing at the central node is transmitted along the outgoing paths as indicated by the detached arrows. The signal returning *via* the self-loop is gx, where g is the branch gain of the self-loop. Since, signals entering the node must add algebraically to give x, it follows that the external signal entering from the left must be (1– g)x. The node and self-loop, therefore, may be replaced by a single branch (b) whose gain is the reciprocal of (1– g). When several branches connect at the node, as in Figure 16(c), it is easyto see that the proper replacement is that shown in Figure 16(d). Quantity g is usuallyreferred to as the loop gain and 1–g is called the loop difference.

Approaching the self-loop effect from another viewpoint, we may treat Figure 16(b) as the residual form of Figure 16(a).This is not, of course, an index-residue. The gain G of (b) is the sum of the gains of all residual paths from the source to the sink in (a). One path passes directly through the node, the second path traverses the loop once before leaving, the third path circles the loop twice, and so on. Hence, the residual gain is given by the infinite geometrical series.

$$G = 1 + g + g^2 + g^3 + ... = \frac{1}{1-g} \tag{6}$$

Which sums to the familiar result. The convergence of this series, for $|g|<l$, posesno dilemma in view of the validity of analytic continuation. The result holds for all values of g except the singular point g = 1, near which the transmission G becomes arbitrarily large.

The self-loop-to-branch transformation places in evidence the basic effect of feed back as a contribution to the denominator of an expression for the gain of a graph in terms of branch gains. In algebra, feedback is associated with division or, more generally, with the inversion of a matrix whose determinant is not identically equal to unity.

THE GENERAL INDEX- RESIDUE OF INDEX ONE

If we restrict attention to a single source and a single sink, then the most general index-residue of index one, or first-index-residue, is that shown in Figure 17(a). Other sources or sinks in the system may be considered separately, without loss of generality, since the system is linear and superposition applies. The knowledge of the self-loop-to branch transformation enables us to write the (source to sink) gain of graph 17(a) by inspection.

The gain is:

$$G = d + \frac{bc}{1-a} \tag{7}$$

When the total index of the graph is greater than one, as in Figure 17(b), it is still a simple matter to find the gain, provided each imbedded feedback unit is only of first index.

For graph 17(b),

$$G = g + \frac{ef}{1-d} + \frac{bcf}{(1-a)(1-d)} \tag{8}$$

With practice, the gain of a graph such as that of Figure 15(a) can be written at a glance, without bothering to make an actual sketch of the residue. The principal source of error lies in the possibility of overlooking a residual path.

Of special interest is the theorem that if each feedback unit in a graph is a simplering of branches, the gain of that graph is equal to the sum of the gains of all open paths from source to sink, each divided by the loop differences of feedback loops encountered by that path. For illustration, we shall apply this theorem to the graph shown in Figure 18. There are nine different open paths from the source to the sink and each one makes contact with the feedback loop.

The resulting gain is:

$$G = \frac{ah + bdh + cgdh + aei + bdei + cgdei + aefj + bdefj + cj}{1 - defg} \tag{9}$$

THE GENERAL INDEX-RESIDUE OF INDEX TWO

Again taking one source and one sink at a time, we shall study the most general second-index-residue shown in Figure 19.

Suppose that, the self-loops are temporarily removed, leaving the simple imbedded ring shown in Figure (b). Graph (b) exhibits five open paths from source to sink, namely i, ab, cd, afd, ceb; and the last four of these encounter the feedback loop ef.

Hence, the gain of graph (b) is:

$$G = i + \frac{ab + cd + afd + ceb}{1 - ef} \tag{10}$$

Now, in order to account for the self-loops g and h in graph 19(a), we need only divide each path gain appearing in expression 10 by the loop difference (1–g) if that path passes through the upper node, and by (1–h) if it passes through the lower node. Paths afd, ceb, and ef, of course, pass through both nodes, and their gains must be dividedby both loop differences. The resulting modification of formula 10 yields the gain of thegeneral second-index-residue.

$$G = i + \frac{\frac{ab}{1-g} + \frac{cd}{1-h} + \frac{afd + ceb}{(1-g)(1-h)}}{1 - \frac{ef}{(1-g)(1-h)}} \tag{11}$$

The derivation of this formula is important only as a demonstration of the power of the method. To find the (source-to-sink) gain of any graph whose feedback units are no worse than second index, we reduce to an index-residue, temporarily remove the selfloops; express the gain as the sum of open path gains, each divided by the loop differences of feedback loops touching that path; and modify the result to account for the originalself-loops.

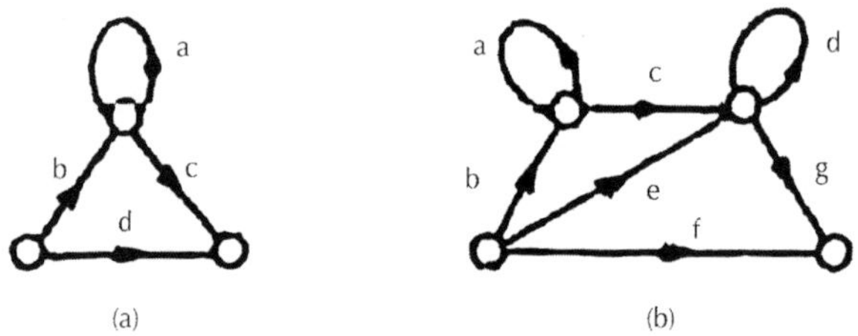

FIGURE 17 Residues having first-index feedback units.

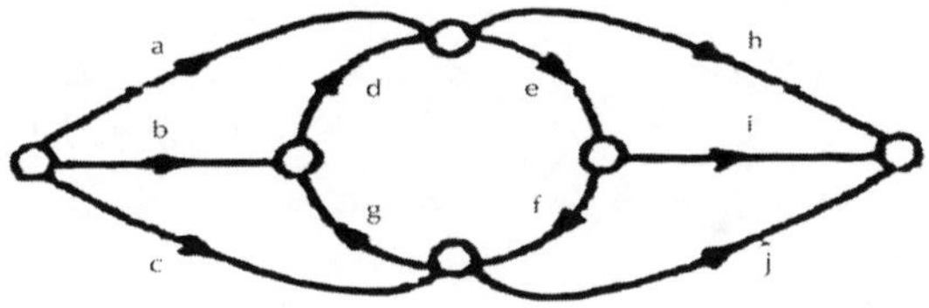

FIGURE 18 A simple ring imbedded in a graph.

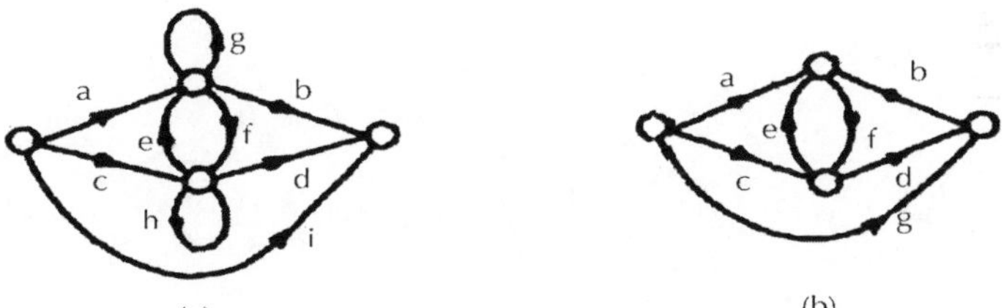

FIGURE 19 The general second-index-residue with and without self-loops.

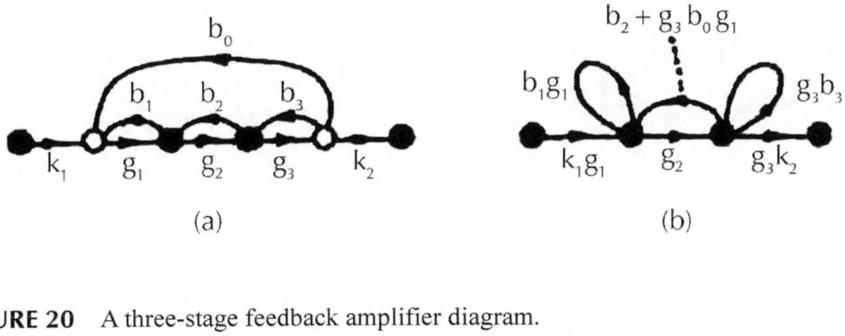

FIGURE 20 A three-stage feedback amplifier diagram.

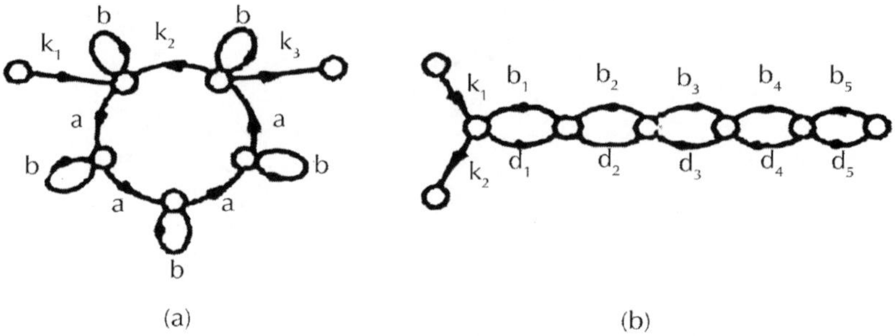

FIGURE 21 Simple high-index structures.

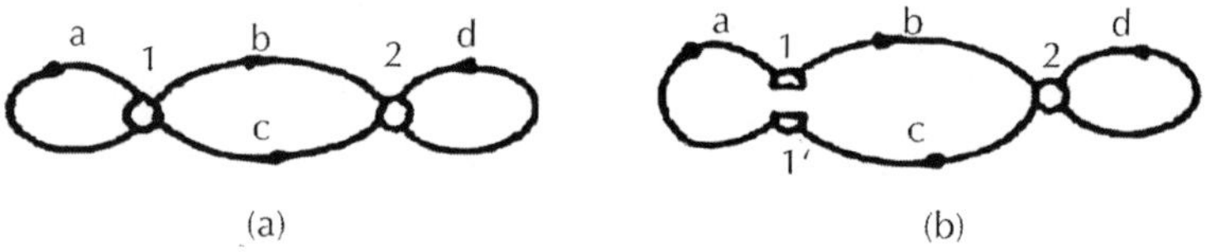

FIGURE 22 The loop gain of a node.

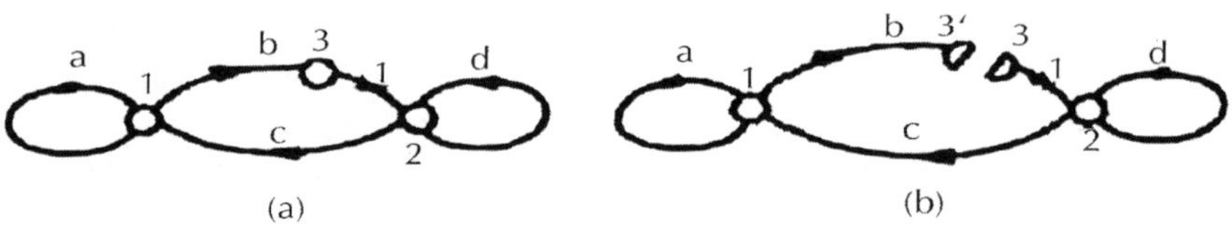

FIGURE 23 The loop gain of a branch.

FIGURE 24 The injection gain at node k.

The importance of the method justifies a final example. Figure 20(a) shows the feedback diagram of a three-stage amplifier having local feedback around each stage and external feedback around the entire amplifier. With the self-loops temporarily removed, the gain of the residue (b) is:

$$G = \frac{k_1 g_1 g_2 g_3 k_2}{1 - g_2 (b_2 + g_3 b_0 g_1)} \tag{12}$$

Since, all paths appearing in expression 12 touch both index nodes, the actual gain of theamplifier is:

$$G = \frac{\dfrac{k_1 k_2 g_1 g_2 g_3}{(1 - b_1 g_1)(1 - b_3 g_3)}}{1 - \dfrac{g_2 (b_2 + b_0 g_1 g_3)}{(1 - b_1 g_1)(1 - b_3 g_3)}} = \frac{k_1 k_2 g_1 g_2 g_3}{(1 - b_1 g_1)(1 - b_3 g_3) - g_2 (b_2 + b_0 g_1 g_3)} \tag{13}$$

GRAPHS OF HIGHER INDEX

The formal reduction process for an arbitrary feedback graph involves a cycle of two steps. First, reduction to an index-residue, and second, replacement of any oneof the self-loops by its equivalent branch. Exactly such cycles are required for reduction to cascade form, where n is the total index of the original graph. Transformation of more than one self-loop at a time is often convenient, even though this may increase the total number of self-loop transformations required in later steps. In practice, of course, the formal procedure should be modified to take advantage of the peculiarities of the structure being reduced. The process effectively ends when the index has been reduced to two, since the evaluation of gain by inspection of the index residue then becomes tractable.

Figure 21 shows two graphs containing high-index feedback units. With self-loops removed from the circular structure (a), the gain is equal to that of the single open forward path $k_1 a^2 k_3$ divided by the loop difference of the closed path $k_2 a^4 a$ and we have:

$$G = \frac{k_1 a^4 k_3}{1 - k_2 a^4} \tag{14}$$

Since, both paths pass through every index node, the reintroduction of the sell-loops yields:

$$G = \frac{\dfrac{k_1 a^4 k_3}{(1-b)^5}}{1 - \dfrac{k_2 a^4}{(1-b)^5}} = \frac{k_1 a^4 k_3}{(1-b)^5 - k_2 a^2} \tag{15}$$

The feedback chain shown in Figure 21(b) is of third index. Instead of reducing it to an index-residue, we shall take advantage of the simplicity of the chain structure to write the gain by a more direct method. First, with the last 4 loops of the chain removed, the gain is:

$$G = \frac{k_1 k_2}{1 - a_1 b_1} \tag{16}$$

Now, the addition of loop a_2b_2 modifies the path gain a_1b_1 to give:

$$G = \frac{k_1 k_2}{1 - \dfrac{a_1 b_1}{1 - a_2 b_2}} \tag{17}$$

Addition of the remaining elements leads to the continued fraction:

$$G = \cfrac{k_1 k_2}{1 - \cfrac{a_1 b_1}{1 - \cfrac{a_2 b_2}{1 - \cfrac{a_3 b_3}{1 - \cfrac{a_4 b_4}{1 - a_5 b_5}}}}} \tag{18}$$

LOOP GAIN AND LOOP DIFFERENCE

Thus, far we have spoken of loop gain only in connection with feedback units of the simple ring type. A more general concept of loop gain will now be introduced. We shall define the loop gain of a node as the gain between the source and sink created by splitting that node. In terms of signal flow, the loop gain of a node is just the signal returned to that node per unit signal transmitted by that node. The loop difference of a node is by definition equal to one minus the loop gain of that node. We shall use the symbol T for loop gains and D for loop differences. In the graph of Figure 22(a), for example, the loop gain of node 1 is equal to the gain from 1 to 1' in graph (b), which shows node 1split into a source 1 and a sink 1'. By inspection:

$$T_1 = a + \frac{bc}{1 - d}. \quad D_1 = 1 - a - \frac{bc}{1 - d}. \tag{19}$$

Another quantity of interest is the loop gain of a branch. Preparatory to its definition, let us replace the branch in question by an equivalent cascade of two branches, whose path gain is the same as the original branch gain. This creates a new node, called an interior node of the branch. The loop gain of a branch may now be defined as the loop gain of an interior node of that branch. To find the loop gain of branch b in Figure 22(a), for instance, we first introduce an interior node 3 as shown in Figure 23(a).The loop gain of branch b is the gain from 3 to 3' in(b),

$$T_{12}\ (\text{or } T_b) = \frac{bc}{(1 - a)(1 - d)} \tag{20}$$

The loop gain of a branch can be designated by either a single or double subscript, whichever is a more convenient specification of the branch. The double subscript is usually preferable, since it avoids confusion with the loop gain of a node. The loop gain of a given node (or branch) evidently involves only the gains of branches which are coupled to that node (or branch). Hence, in computing T, we need to consider only the feedback unit containing the node (or branch) of interest.

Having defined the loop gain of a node, we may extend the simple self-loop equivalence to a more general form which may be stated as follows. If an external signal $x_0$is injected into node k of a flow graph,as shown in Figure 24, the injection gain from the external source to node k is:

$$G_k = \frac{x_k}{x_0} = \frac{1}{1\, T_k} = \frac{1}{D_k} \tag{21}$$

The very nature of the reduction process for an arbitrary (finite) graph implies that the gain is a rational function of the branch gains. In other words, the gain can always be expressed as a fraction whose numerator and denominator are each algebraic sums of various branch gain products. Moreover, the gain G is a linear rational function of any one of the branch gains g.

Thus:

$$G = \frac{ag + b}{cg + d} \tag{22}$$

Where quantities a, b, c, and d are made up of other branch gains. To prove this we may insert two interior nodes into the specified branch g, as shown in Figure 25(a) and (b), and then consider the residue(c), which contains only the source, the sink, and the two interior nodes.The gain of this residue evidently can beexpressed as a linear-rational function of g. It is also apparent that if branch g is directly connected to either the source or the sink, or to both, then the source-to-sink gain G is a linear function of the branch gain g, that is:

$$G = ag + b \tag{23}$$

where a and b depend upon other branch gains.

The foregoing results apply equally well to loop gains and loop differences, sinceT and D, by their definitions, have the character of gains. Any loop difference D_k is a rational function of the branch gains, a linear rational function of any single branch gain, and a linear function of the gain of any branch connected directly to node k.

We shall now derive an important fundamental property of loop differences which is of general interest. Consider an arbitrary graph containing nodes 1, 2, 3, ..., n, and let nodes m + l, m + 2, ... , n–1, n be removed, to gether with their connecting branches, so that only nodes 1, 2, 3, ... , m remain.Now suppose that the graph is reduced to a residue showing only nodes m–1, and m, as in Figure 26. Branches a, b, c, and d account for all coupling among nodes 1, 2, 3, ..., m of the original graph. Sources and sinks may beignored, ofcourse,since, only feedback branches are of interest in loop difference calculations.

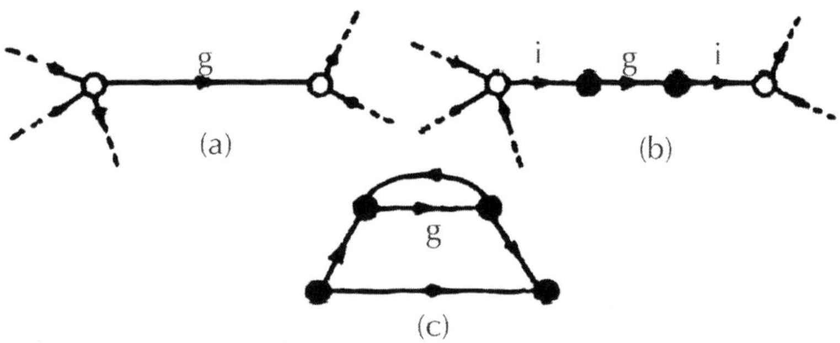

FIGURE 25 The graph gain as a function of a particular branch gain.

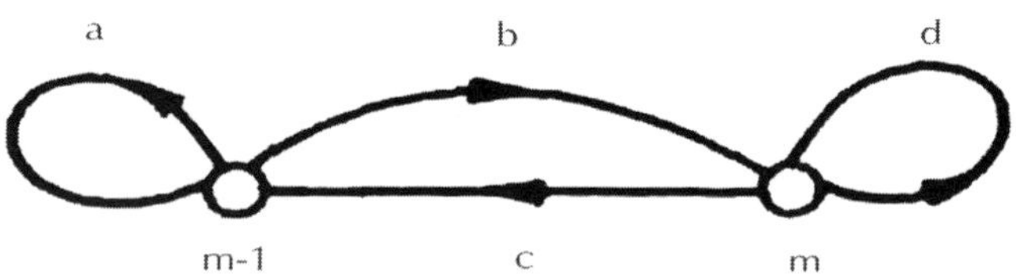

FIGURE 26 A residue showing nodes m–1 and m.

Let us define the partial loop difference D'_k as the loop difference of node k with-only the first k nodes taken into account.

By inspection of Figure 2:

$$D'_m = 1 - d - \frac{bc}{1 - a} \tag{24}$$

$$D'_{m-1} = 1 - a \tag{25}$$

and

$$D'_{m-1} D'_m = (1 - a)(1 - d) - bc. \tag{26}$$

If the numbers of nodes m–1 and m are interchange in Figure 26, then:

$$D'_m = 1 - a - \frac{bc}{1 - d} \tag{27}$$

$$D'_{m-1} = 1 - d \tag{28}$$

and the product given in Equation 26 is unaltered. Since, this result holds for any value of m, and since, a sequence may be transformed into any other sequence by repeated adjacent interchanges (1234 can become 4321, for example, by the adjacent interchanges 1243, 2143, 2413, 4213, 4231, 4321), it follows that the product:

$$\Delta'_m = D'_1\ D'_2\ D'_3 \ldots D'_{m-1} D'_m \tag{29}$$

is independent of the order in which the first m nodes are numbered.With all n nodes present, we have $D'_n = D_n$ and:

$$Ä = D'_1 D'_2 D'_3 \ldots D'_{n-1} D'_n \tag{30}$$

QuantityΔ, which we shall call the determinant of the graph, is invariant for any order of node numbering. Equation 30 shows that the determinant of any graph is the product of the determinants of its imbedded feedback units, and that the determinant of a cascade graph is unity.

The dependence of Δ upon the branch gains may be deduced as follows. Let g be any branch directly connected to node n, w hence it follows that D_n is a linear function of branch gain g and that the partial loop differences D'_k are independent of g.

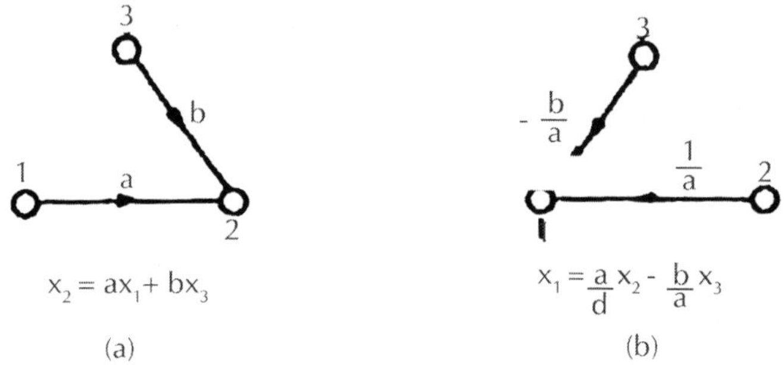

FIGURE 27 Branch inversion in a linear graph.

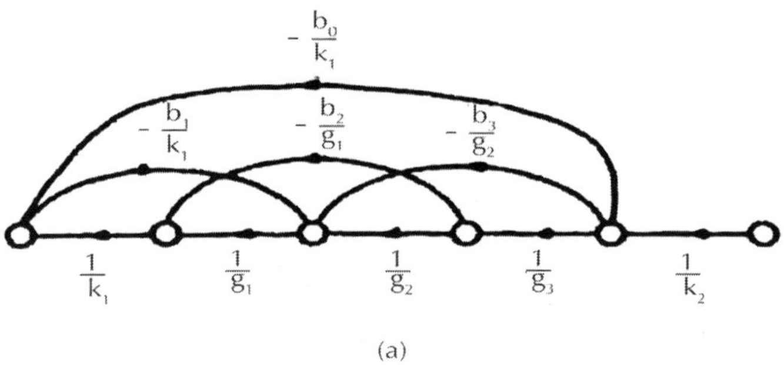

FIGURE 28 The result of path inversion in Figure 20(a).

Hence, Δ is a linear function of g. Since the numbering of nodes is arbitrary, Δ must be a linear function of any given branch gain in the graph. The determinant Δ, therefore, is composed of an algebraic sum of products of branch gains, with no branch gain appearing more than once in a single product.

From Equation 29 and Equation 30 we see that D_n is the ratio of Δ to Δ'_{n-1}. Since, the node number is arbitrary,we may write:

$$D_k = \frac{\Delta}{\Delta_k} \tag{31}$$

where Δ_k is to be computed with node k removed. Once Δ is expressed in terms of branch gains, Δ_{kk} may be found by nullifying the gains of branches connected to node k.

The introduction of an interior node into any branch leaves the value of Δ unaltered. To prove this we may number the new node zero, when $D'_0 = 1$ and the other partial loop differences are unchanged. It follows directly that the loop difference of any branch jk is given by equation.

$$D_{jk} = \frac{\Delta}{\Delta_{jk}} \tag{32}$$

Where Δ_{jk} is to be computed with branch jk removed, that is, with $g_{jk} = 0$.

Incidentally, if we write the linear equations associated with the flow graph and then evaluate the injection gain G_k by Kramer's rule (that is, by inverting the matrix of the equations), we find from Equation 21 and Equation 31 that Δ is just the value of the determinant of these equations.

INVERSE GAINS

We have already seen how the form of a flow graph is altered by the inversion of a path.For linear graphs it is profitable to continue with an inquiry into the quantitative effects of inversion. Figure 27(a) shows 2 branches which may be imagined to form-part of a larger graph. The signal entering node 2 *via* branch b is bx_3. The contribution arriving from branch a, then, must be x_2-bx_3, since, the sum of these two contributions is equal to x_2. Hence, given x_2 and x_3, the required value of $x_1$is that indicated in graph (b).

The general scheme is readily apparent andmay bestated as follows.The inversion of any branch jk is accomplished by reversing that branch and inverting its gain, and shifting any other branch ik having the same nose location k to the new position ij and dividing its gain by the negative of the original branch gain g_{jk}.

For gain calculations, the usefulness of inversion lies in the fact that the inversionof a source-to-sink path yields a new graph whose source-to-sinkgain is the inverse of the original source-to-sink gain.Since, inversion may accomplish a reduction of index, the inverse gain may be much easier to find by inspection. For illustration, we shall invert path $k_1g_1g_2g_3k_2$ in Figure 20(a) to obtain the graph shown in Figure 28. The newgraph is a cascade structure of zero index.By inspection of the new graph, the inverse gain of the original graph is:

$$\frac{1}{G} = \frac{1}{K_2}\left[\left(\frac{1}{g_3 g_2} - \frac{b_3}{g_2}\right)\left(\frac{1}{g_1 k_1} - \frac{b_1}{k_1}\right) - \frac{b_2}{g_3 g_1 k_1} - \frac{b_0}{k_1}\right]. \tag{33}$$

Simplification yields:

$$\frac{1}{G}=\frac{1}{K_1K_2}\left[\frac{1}{g_2}\left(\frac{1}{g_1}-b_1\right)\left(\frac{1}{g_3}-b_3\right)-\frac{b_2}{g_1g_3}-b_0\right] \tag{34}$$

which proves to be identical with Equation 13.

A simpler example is offered by Figure 21(a).Inversion of the open source-to-sink path gives the structure shown in Figure 29. By inspection ofthe new graph, we find:

$$\frac{1}{G}=\frac{1}{k_3}\left[\left(\frac{1}{a}-\frac{b}{a}\right)^4\left(\frac{1}{k_1}-\frac{b}{k_1}\right)-\frac{k_2}{k_1}\right]=\frac{(1-b)^5}{k_1k_3a^4}-\frac{k_2}{k_1k_3} \tag{35}$$

which checks Equation 15.

NORMALIZATION

In the general analysis of an electrical network, it is often convenient to alterthe impedance level or the frequency scale by a suitable transformation of element values. A similar normalization sometimes proves useful for linear flow graph analysis. The self-evident normalization rule may be stated as follows. If each branch gain g_{jk} ismultiplied bya scale factor f_{jk}, with the scale factors so chosen that the gains of all closed paths are unaltered, then the gain of the graph is multiplied by$f_{12}f_{23}.....f_{nm}$, where 1, 2, 3, ... m, n is any path from the source 1 to the sink n.

Figure 30 illustrates a typical normalization. Graph (a) might represent a two-stage amplifier with isolation between the two stages, local feedback around each stage, and external feedback around both stages. The normalization shown in (b) brings out very clearly the fact that certain branch gains may be taken as unity without loss of generality.

ILLUSTRATIVEAPPLICATIONS OFFLOW GRAPH TECHNIQUES

The usefulness of flow graph techniques for the solution of practical analysis problems is limited by two factors—our ability to represent the physical problem in the form of a suitable graph, and our facility in manipulating the graph. The first factor has not yet been considered. We turn to it now with the necessary background material at hand.

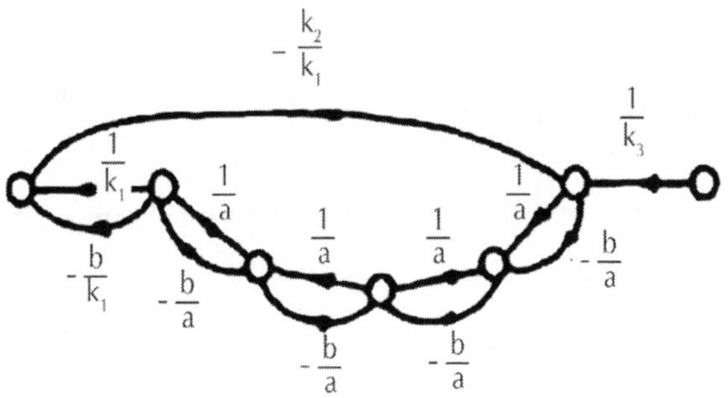

FIGURE 29 The result of path inversion in Figure 21(a).

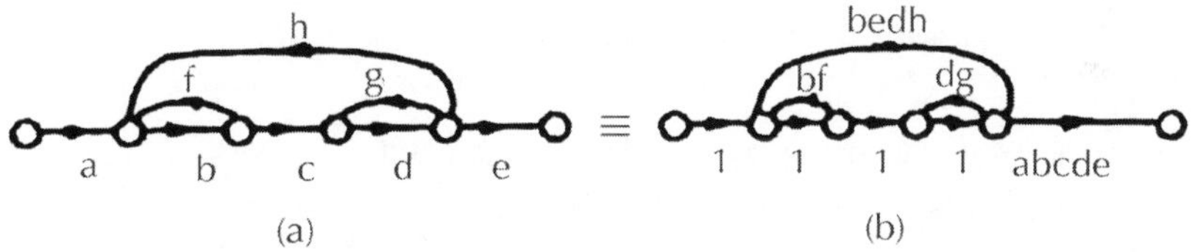

FIGURE 30 Normalization.

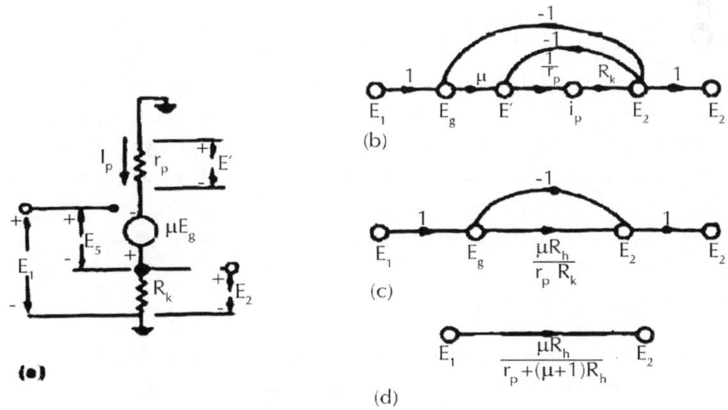

FIGURE 31 Flow graphs for a cathode follower.

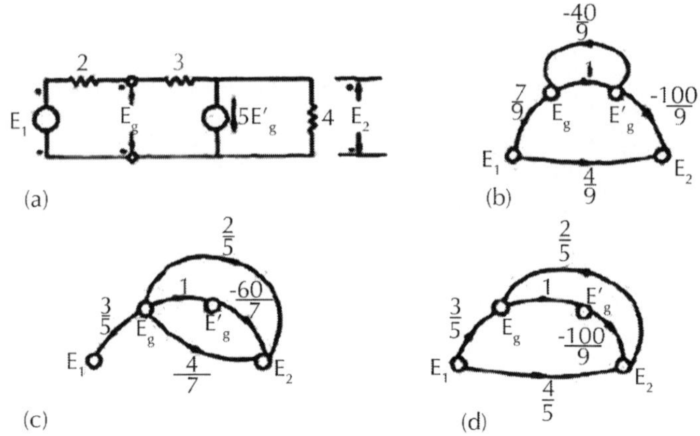

FIGURE 32 Anamplifier with grid-to-plate impedance.

The process of constructing a graph is one of tracing a succession of causes and effects through the physical system. One variable is expressed as an explicit effect due to certain causes they, inturn, are recognized as effects due to still other causes. In order to be associated with a single node, each variable must play a dependent role only once. A link in the chain of dependency is limited in extent only by perception of the problem. The formulation may be executed in a few complicated steps or it may be subdivided into a larger number of simple ones, depending upon our judgment and knowledge of the particular system under consideration. No specific rules can be given for the best approach to ananalysis problem.There in lies the challenge and the possibility of an elegant solution. Whatever the approach, flowgraphs offer a structural visualization of the interrelations among the chosen variables. It is quite possible, of course, to construct an incorrect graph just as it is entirely possible to write a set of equations which do not properly represent the physical problem. The direct formulation of a flow graph from a physical problem without actually writing the chosen equations requires some practice before confidence is gained. It is hoped that the following examples, taken mostly from electronic circuit analysis, willbe suggestive.

VOLTAGE GAIN CALCULATIONS

Figure 31(a) shows the low-frequency linear in cremental equivalent circuit of a cathode follower. Suppose that we want to find the gain E_2/E_1 in terms of the circuit constants. Proceeding very cautiously insmall steps, we might construct the graph shown in Figure3l(b). This graph states that $E_g = E_1 - E_2$, $E' = \mu E_g - E_2$, $I_p = E'/r_p$, and $E_2 = R_k I_p$. Alternatively, were we able to recognize at the out set the direct dependence of E_2 upon E_g, then graph 3l(c) could have been sketched by inspection of the circuit.The more extensive our powers of perception, the simpler the formulation. Powerful perception (or a familiarity with the cathode follower) would permit us to construct graph 3l(d) directly from the network shown in Figure 3l(a). The reader is invited to evaluate the gains of graphs 3l(b) and (c) by inspection and to compare them with Figure 3l(d).

Another example is offered by the amplifier of Figure 32(a). For convenience of illustration, the impedances and the trans conductance have been given numerical values. In this circuit the grid voltage influences the output voltage both by transconductance action and by direct coupling through the grid-to-plate impedance. To avoid confusion between the actual voltage E_g and the factor E_g appearing in the transconductance current, it is very helpful to designate one of them with a prime while we are setting up the graph. This distinction splits node E_g. It is a simple matter to complete the graph with a unity-gain branch representing the equation $E'_g = E_g$, which effectively rejoins the node.

The direct application of superposition, with voltage E_1 and current SE'_g treated as independent electrical sources, each influencing the dependent quantities E_g and E_2, leads to graph (b) of Figure 32. The gain from E'_g to E_g, for example, is the product of a transconductance 5, a current division ratio 4/9, and an impedance 2, as measured with $E_1 = 0$.

An alternative approach, actually equivalent to classical network formulation on the electrical-node-pair-voltage basis, gives graph 32(c). Here, E_2 is expressed as a function of E_g and E'_g. In accordance with superposition, the gain from E'_g to E_2 must be computed with $E_g = 0$ (rather than $E_1 = 0$, as in the previous graph). Hence, in this particular calculation, the indepedance presented to the current source does not include element 2. The other independent electrical-node-pair voltage E_g is expressed interms of E_1 and E_2, as shown.

Graph 32(d), a third possibility, is actually the simplest and most elegant of the three. Responding toa certain physical appeal, we express E_2 interms of the two electrical sources, as in graph 32(b). Taking advantage of the fact that E_2 and $5E'_g$ are across the same electrical node-pair, we formulate E_g in terms of E_1 and E_2 as in graph 32(c). This has topological appeal, since the resulting feedback loop touches both open paths from E_1 to E_2. As a result, the graph gain is a simple fractional function of the branch gains. The verification of graphs (b), (c), and (d) of Figure 32 and the evaluation of their gains is suggested as an exercise for the reader. The answer is –8/7.

If symbols are substituted for the numerical element values in the circuit, the suitability of the structure of Figure 32(d) for this particular problem becomes more apparent.

THEIMPEDANCE FORMULA

Suppose, that the input or output impedance Z of an electronic circuit is influenced by a certain tube transconductance in such a manner that the effect is not immediately obvious.To find Z we must introduce a set of variables and write the equations relating them. Let us choose the terminal current and voltage, I and E = IZ, to gether with the grid voltage E_g of the offending tube, as shown in Figure 33(a).

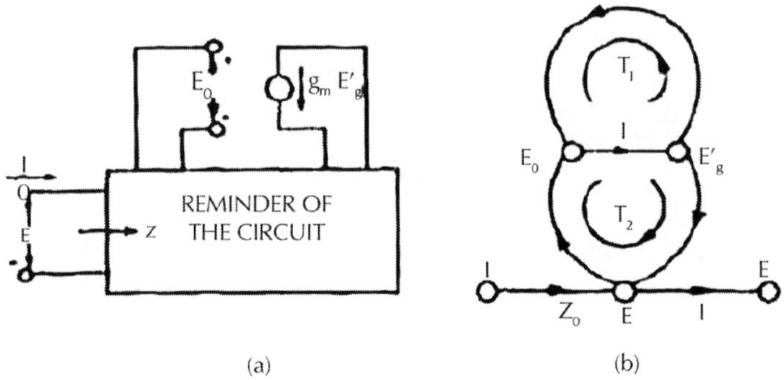

FIGURE 33 The circuit and graph for terminal impedance formulation.

The graphical structure which naturally suggests itself, perhaps, is that of the previous problem, Figure 32(b), with a source I and a sink E. Since, E and I are located at the same pair of terminals, however, it is just as easy to express E_g interms of E'_g and E, rather than E'_g and I. This choice gives graph (b) of Figure 33, which is particularly convenient for present purpose. Notice that the structure of Figure 33(b) is obtainable directly from that of Figure 32(b) by inversion of the source-to-sink branch.

The three gains of interest in Figure 33(b) are:

$$Z_0 = \left(\frac{E}{I}\right)_{E'_g=0} = \text{the impedence without feedback} \tag{36}$$

$$T_g^{sc} = \left(\frac{E_g}{E'_g}\right)_{E=0} = \text{the short-circuit loop gain} = T_1 \tag{37}$$

$$T_g^{oc} = \left(\frac{E_g}{E'_g}\right)_{I=0} = \text{the open-circuit loop gain} = T_1 + T_2. \tag{38}$$

The terminal impedance is given by the graph gain:

$$Z = \frac{Z_0}{1 - \frac{T_2}{1 - T_1}} = Z_0\left(\frac{1-T_1}{1-T_1-T_2}\right) \tag{39}$$

which may be identified as the well-known feedback formula:

$$Z = Z_0\left(\frac{1-T_g^{sc}}{1-T_g^{oc}}\right). \tag{40}$$

The conclusion is that flow graph methods providea relatively un cluttered derivation of this classical result.

Flow graph representation also brings out the similarities between feedback formulas for electronic circuits and compensation theorems for passive networks. Consider, for comparison, the determination of the input impedance of the circuit shownin Figure 34(a).

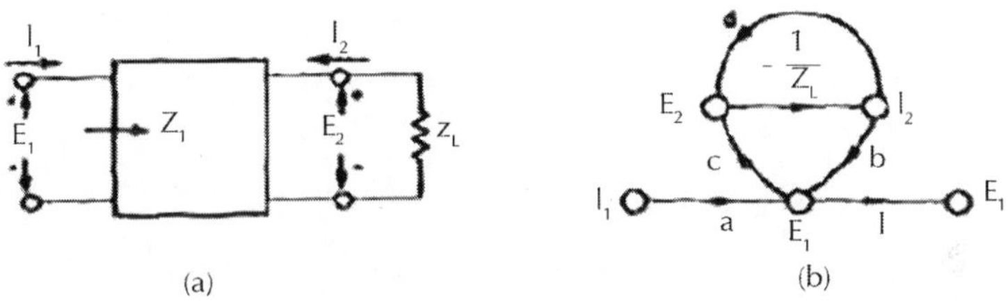

FIGURE 34 The effect of load impedance upon input impedance.

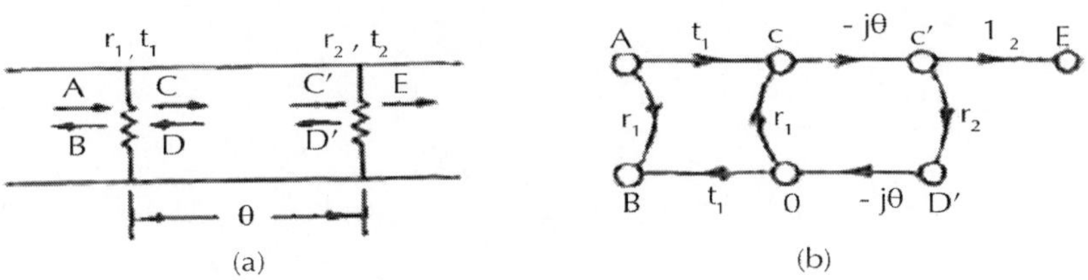

FIGURE 35 Two discontinuities on a transmission line.

Superposition tells us that the branch gains of the accompanying graph, Figure 34(b), have the physical interpretations.

$$Z_1^{oc} = \left(\frac{E_1}{I_1}\right)_{I_2=0} = \text{open-circuit input impedence} = a \tag{41}$$

$$Z_2^{oc} = \left(\frac{E_2}{I_2}\right)_{I_1=0} = \text{open-circuit output impedence} = bc + d \tag{42}$$

$$Z_2^{sc} = \left(\frac{E_2}{I_2}\right)_{E_1=0} = \text{short-circuit output impedence} = d \tag{43}$$

By analogy with the previous problem:

$$Z_1 = Z_1^{oc} \frac{1 + \dfrac{Z_2^{sc}}{Z_L}}{1 + \dfrac{Z_2^{oc}}{Z_L}} = Z_1^{oc}\left(\frac{Z_L + Z_2^{sc}}{Z_L + Z_2^{oc}}\right). \tag{44}$$

AWAVE REFLECTION PROBLEM

The transmission line shown in Figure 35(a) has two shunt discontinuities spaced θ electrical radians apart. A voltage wave of complex amplitude A is incident upon the first discontinuity from the left. We desire to find the resulting reflection B and the transmitted wave E. Let C, D, C', D' be the waves traveling in opposite directions just to the right of the first obstacle and just to the left of the second. In addition, let r and t denote the per unit reflection or transmission of a single discontinuity.

The accompanying graph 35(b) is self-explanatory. The only feedback loop present is the simplering CC'D'DC. By inspection of this graph, the over-all reflection and transmission coefficients are:

$$\frac{B}{A} = r_1 + \frac{t_1^2 r_2 e^{-j2\theta}}{1 - r_1 r_2 e^{-j2\theta}} \tag{45}$$

$$\frac{E}{A} = \frac{t_1 t_2 e^{-j2\theta}}{1 - r_1 r_2 e^{-j2\theta}}. \tag{46}$$

A LIMITERDESIGN PROBLEM

Figure 36(a) shows a vacuum-tube circuitcommonly employed as a two-way limiter or level selector. The static transfer curve shown in Figure 36(b) exhibits a high-gain central region limited on each side by cut off. In the neighborhood of point p, where both tubes are conducting, the linear in crement circuit of Figure 36(c) applies. If we design the incremental circuit for infinite gain, then the transfercurve becomes vertical at point p, and the switching interval is made desirably small.

Assume for simplicity that the voltage divider feeding the second grid has a resistance much greater than R_1 (or let R_1 denote the combined parallel resistance). Now let us attempt to formulate E_1 in terms of E_0 and E_k by superposition. With $E_k = 0$, the ratio E_1/E_0 is simply the gain of a grounded-cathode stage. Similarly, with $E_0 = 0$, the first tube becomes a grounded-grid stage driven by E_k. This gives us branches 01 and kl in the flow graph shown in Figure 36(d). Branches 12 and k_2 follow the same pattern for the second tube.We must now formulate $E_k$in a convenient manner. One possibility is the computation of thetwo tube currents-E_1/R_1 and $-E_2/R_2$, whose sum may be multiplied by $R_k$to obtain E_k, as shown.

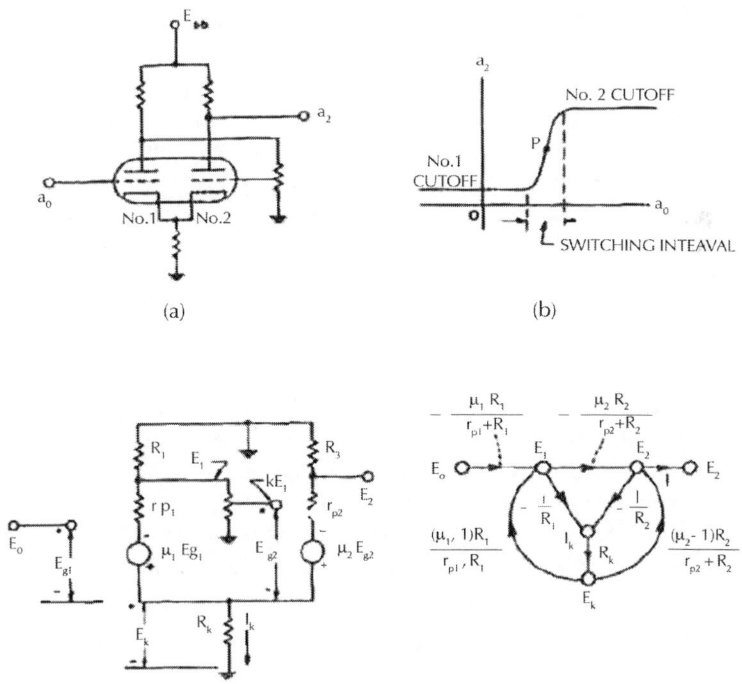

FIGURE 36 A cathode-coupled limiter.

The resulting graph is of index one, and either E_k or $I_k$may betaken as the index node. The index-residue would have the familiar form shown in Figure 17(a). For infinite gain we need only specify that the loop gain of node E_k(or nodeI_k or branch R_k) must be unity. By inspection of the graph, the three paths entering T_k are k12k, k1k, and k2k.

Hence:

$$T_k = R_k \left[\frac{k(\mu_1 + 1)\mu_2 R_1}{(r_{p1} + R_1)(r_{p2} + R_2)} - \frac{\mu_1 + 1}{r_{p1} + R_1} - \frac{\mu_2 + 1}{r_{p2} + R_2} \right] = 1. \tag{47}$$

It is a simple matter to solve this equation for the desired value of the voltage divider parameter k.

CONCLUDING REMARKS

The flow graph offers a visual structure, a universal graphical language, acommon ground upon which causal relationships among a number of variables may be laid outand compared. From this view point the similarity between two physical problems arises not from the arrangement of physical elements or the dimensions of the variables but rather from the structure of the set of relationships which we care to write.

The organization of the problem comes from within minds and feedback is present only if we perceive a closed chain of dependency. The challenge facing us at the start of an analysis problem is to express the pertinent relationships as a meaningful and elegant flowgraph.The topological properties of the graph may then be exploited in the manipulations and reductions leading to a solution.

FEEDBACK THEORY: FURTHER PROPERTIES OF SIGNAL FLOW GRAPHS

BACKGROUND

There are many different paths to the solution of a set of linear equations. The formal method involves inversion of a matrix. We know, however, that there are many different ways of inverting a matrix: Determinantal expansion in minors, systematic reduction of a matrix to diagonal form, partitioning into submatrices, and so forth, each of which has its particular interpretation as a sequence of algebraic manipulations within the original equations. A determinantal expansion of specialinterest is:

$$D = \sum a_{1i} a_{2j} a_{3k} \ldots . a_{nz} \tag{1}$$

where a_{mp}..........,.is the element in the *m*th row and *p*th column of a determinant having *n* rows, and the summation is taken overall possible permutations of the column subscripts. (The sign of each term is positive or negative in accord with an even or odd number of successive adjacent column-subscript interchanges required to produce a given permutation.) Since, the solution of a set of linear equations involves ratios of determinantal quantities, (1) suggests the general idea that a linear system analysis problem should be interpretable as a search for all possible combinations of something or other, and that the solution should take the form of a sum of products of the somethings, whatever they are, divided by another such sum of products. Hence, instead of undertaking a sequence of operations, we can find the solution by looking for certain combinations of things. The method will be especially use fulif these combinations have a simple interpretation in the context of the problem.

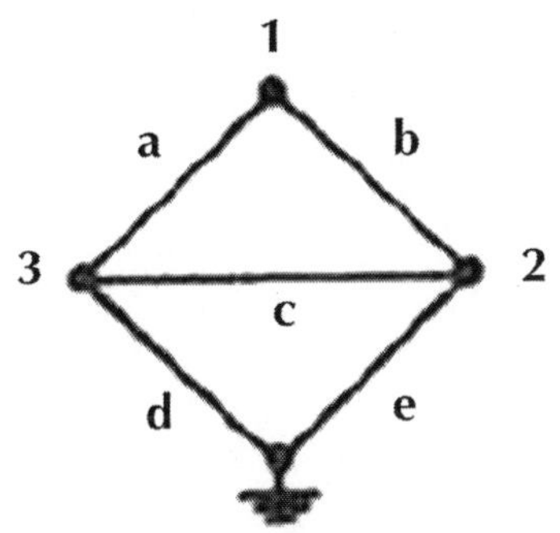

FIGURE 1 An electrical network graph.

As a concrete illustration of the idea, consider the electrical network graph shown in Figure 1. For simplicity, let the branch admittances be denoted by letters a, b, c, d, and e. This particular graph has three independent node pairs. Locate all possible sets of three branches which do not contain closed loops and write the sum of their branch admittance products as the denominator of Equation(2).

$$Z_{12} = \frac{ab + ac + bc + bd}{abd + abe + acd + ace + ade + bcd + bce + bde} \tag{2}$$

Now, locate all sets of two branches which do not form *closed loops* and which also *do not contain anypaths from node 1 to ground or from node 2 to ground.* Write the sum of their branch admittance products as the numerator of expression (2).The result is the transfer impedance between nodes1 and 2, that is, the voltage at node2 when a unit current is injected at node 1. Any impedance of a branch network can be found by this process (Ku, 1952).

So much for electrical network graphs. The main concern in this paper is withsignalflow graphs, (Mason, 1953) whose branches aredirected. Tustin (Mason, 1953) has suggested that the feedback factor for a flow graph of the form shown in Figure 2 can be formulated by combining the feedback loopgains in a certain way.

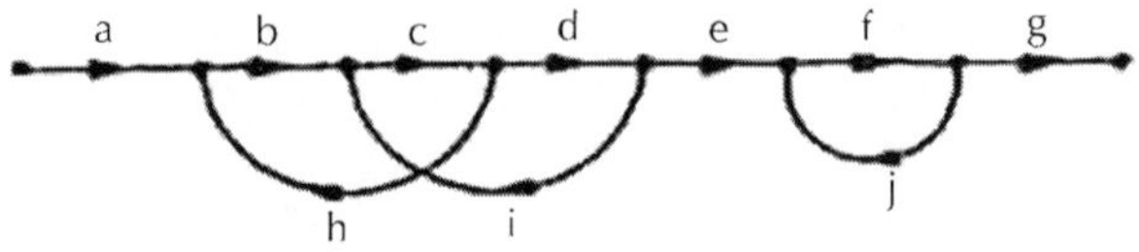

FIGURE 2 The flow graph of an automatic control system.

The three loop gains are:

$$T_1 = bch \tag{3a}$$

$$T_2 = cdi \tag{3b}$$

$$T_3 = fj \tag{3c}$$

and the forward path gain is:

$$G_0 = abcdefg. \tag{4}$$

The gain of the complete system is found to be:

$$G = \frac{G_0}{\left[1 - (T_1 + T_2)\right](1 - T_2)} \tag{5}$$

and expansion of the denominator yields:

$$G = \frac{G_0}{1 - (T_1 + T_2 + T_3) + (T_1 T_3 + T_2 T_3)}. \tag{6}$$

Tustin recognized the denominator as unity plus the sum of all possible products of loop gains take no neat a time $(T_1+T_2+T_3)$, two at a time, $(T_1T_2+T_2T_3)$, three at a time, and so forth, excluding products of loops that touch orpartially coincide. The products $T_1T_2$and $T_1T_2T_3$ are properly and accordingly missing in this particular example. The algebraic sign alternates, as shown, with each succeeding group of products.

Tustin did not take up the general case but gave a hint that a graph having several different forward paths could be handled by considering each path separately and superposing the effects. Detailed examination of the general problem shows, infact, that the form of Equation(6) must be modified to include possible feedback factors in the numerator. Otherwise Equation (6) applies only to those graphs in which each loop-touches all forward paths.

The purposes of this appendix are to extend the method to a general form applic able to any flowgraph to present a proof of the general result, and to illustrate the usefulness of such flow graph techniques by application to practical linear analysis-problems. The proof will be given last. It is tempting to add, at this point, that a better understanding of linear analysis is a great aid in problems of nonlinear analysis and-linear or nonlinear design.

A BRIEFSTATEMENT OF SOMEELEMENTARY PROPERTIES OFLINEAR SIGNAL FLOWGRAPHS

A signal flow graph is a network of directed branches which connect at nodes. Branch *jk* originates at node *j* and terminates upon node *k*, the direction from *j* to *k* being indicated by an arrow head on the branch. Each branch *jk* has associated with it a quantity called the branch gain g_{jk} and each node *j* has an associated quantity called the node signal x_j. The various node signals are related by the associated equations:

$$\sum_i x_j g_{jk} = x_k, \qquad k = 1, 2, 3, \ldots\ldots \tag{7}$$

The graph shown in Figure 3, for example, has equations:

$$ax_1 + dx_3 = x_2 \tag{8a}$$

$$bx_2 + fx_4 = x_3 \tag{8b}$$

$$ex_2 + cx_3 = x_4 \tag{8c}$$

$$gx_3 + hx_4 = x_5 \tag{8d}$$

We shall need certain definitions. A *source* is a node having only outgoing branches (node1 in Figure3). A sink is anode having only incoming branches. A*path* is any continuous succession of branches traversed in the indicated branch directions. A*forward Path* is a path from source to sink along which no node is encountered more than once(*abch,aeh,aejg,abg*, in Figure 3).

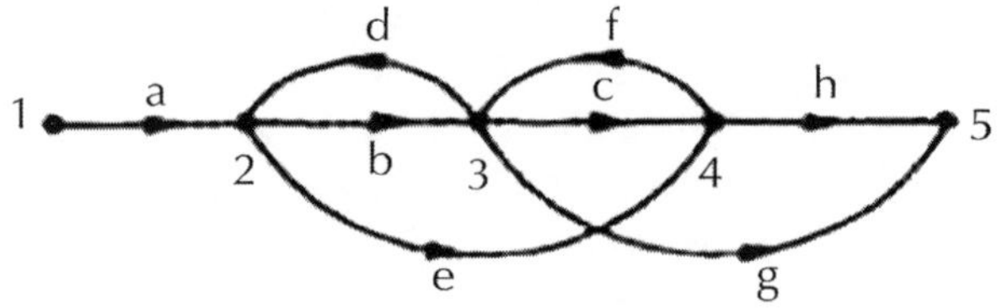

FIGURE 3 A simple signal flow graph.

A *feedback loop* is a path that forms a closed cycle along which each node is encountered onceper cycle (*bd*, *cf*, *def*, but not *bcfd*, in Figure 3). A *path gain* is the product of the branch gains along that path. The *loopgain* of a feedback loop is the product of the gains of the branches forming that loop. The *gain of a flow graph* is the signal appearing at the sink perunit signal applied at the source. Only one source and one sink need be considered, since sources are superposable and sinks are independent of each other.

Additional terminology will be introduced as needed.

GENERAL FORMULATION OFFLOW GRAPH GAIN

To begin with an example, consider the graph shown in Figure 4. This graph exhibits three feedback loops, whose gains are:

$$T_1 = h \tag{9a}$$

$$T_2 = fg \tag{9b}$$

$$T_3 = de \tag{9c}$$

and two forward paths, whose gains are:

$$G_1 = ab \tag{10a}$$

$$G_2 = ceb. \tag{10b}$$

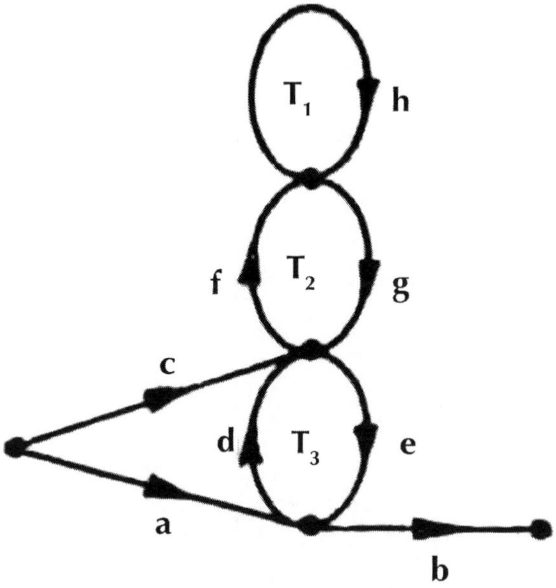

FIGURE 4 A flow graph with three feedback loops.

To find the graph gain, first locate all possible *sets of non-touching loops* andwrite the algebraic sum of their gain products as the denominator of (11).

$$G = \frac{G_1(1 - T_1 - T_2) + G_2(1 - T_1)}{1 - T_1 - T_2 - T_3 + T_1T_3} \tag{11}$$

Each term of the denominator is the gain product of aset of non-touching loops. The algebraic sign of the termis plus (or minus) for an even (or odd) number of loops in the set. The graph of Figure 4 has no sets of three or more non-touching loops. Taking the loops two at a time we find only one permissible set, T_1T_3. When the loops are taken one at a time the question of touching does not arise, so that each loop in the graph is itself an admissible "set." For completeness of form we may also consider the set of loops taken "none at a time" and, by analogy with the zero's power of a number, interpret its gain product as the unity termin the denominator of Equation (11). The numerator contains the sum of all forward pathgains,eachmultiplied bya factor. The factor for a given forward path is made up of all possible sets of loops *which do not touch each other* and which also do *not touch that forward path.*The first forward path(G_1 = *ab*)t ouches the third loop, and T_3 is therefore, absent from the first numerator factor. Since, the secondpath (G_2=*ceb*) touches both T_2 andT_3, onlyT_1enters the second factor.

The general expression for graph gain may be written as:

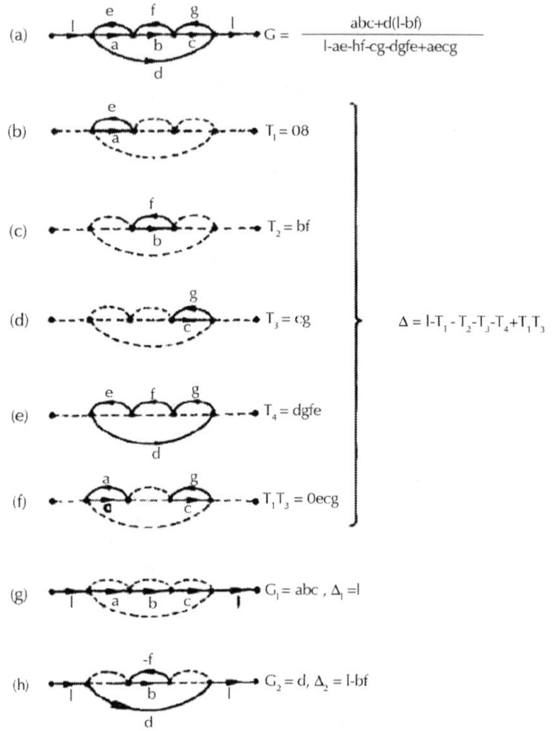

FIGURE 5 Identification of paths and loop sets.

$$G = \frac{\sum_k G_k \Delta_k}{\Delta} \tag{12a}$$

Wherein:

G_k = gain of the kth forward path (12b)

$$\Delta = 1 - \sum_m P_{m1} + \sum_m P_{m2} - \sum_m P_{m3} + \dots \tag{12c}$$

P_{mr} = gain productof the mth possible combination of r non-touching loop (12d)

Δ_k = the value of Δ for that part of the graph not touching the kth forward path. (12e)

The form of Equation (12a) suggests that we call Δ the determinant of the graph, and call Δ_k the *cofactor* of forward path *k*.

A subsidiary result of some interest has to do with graphs whose feedback loops form non-touching subgraphs. To find the *loop subgraph* of any flow graph, simply remove all of those branches *not* lying in feedback loops, leaving all of the feedback loops, and nothing but the feedback loops. In general, the loop sub graph may have a number of non-touching parts. The useful fact is that *the determinant of acomplete flowgraph is equal to the product of the determinants of each of the non-touchingparts in its loop subgraph.*

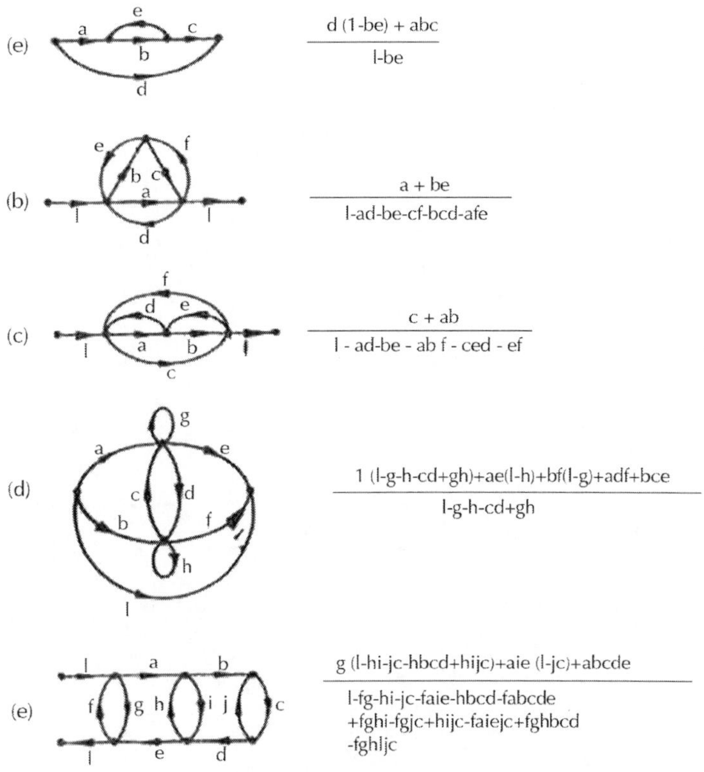

FIGURE 6 Sample flow graphsandtheir gain expressions.

ILLUSTRATIVE EXAMPLES OF GAIN EVALUATION BY INSPECTION OF PATHS AND LOOP SETS

Equation (12) is formidable at first sight but the idea is simple. More examples will help illustrate its simplicity. Figure 5 shows the first of these displayed in minute detail—(a) the graph to be solved, (b)–(f) the loop sets contributing to Δ, (g) and (h) the forward paths and their cofactors. Figure 6 gives several additional examples onwhich you may wish to practice evaluating gains by inspection.

ILLUSTRATIVE APPLICATIONS OF FLOW GRAPH TECHNIQUES TO PRACTICAL ANALYSIS PROBLEMS

The study of flow graphs is a fascinating topological game and therefore, from one view point, worthwhile in its own right. Since, the associated equations of a linear flow graph are in cause-and-effect form, each variable expressed explicitly in terms of others, and since, physical problems are often very conveniently formulated in just this form, the study of flow graphs assumes practical significance.

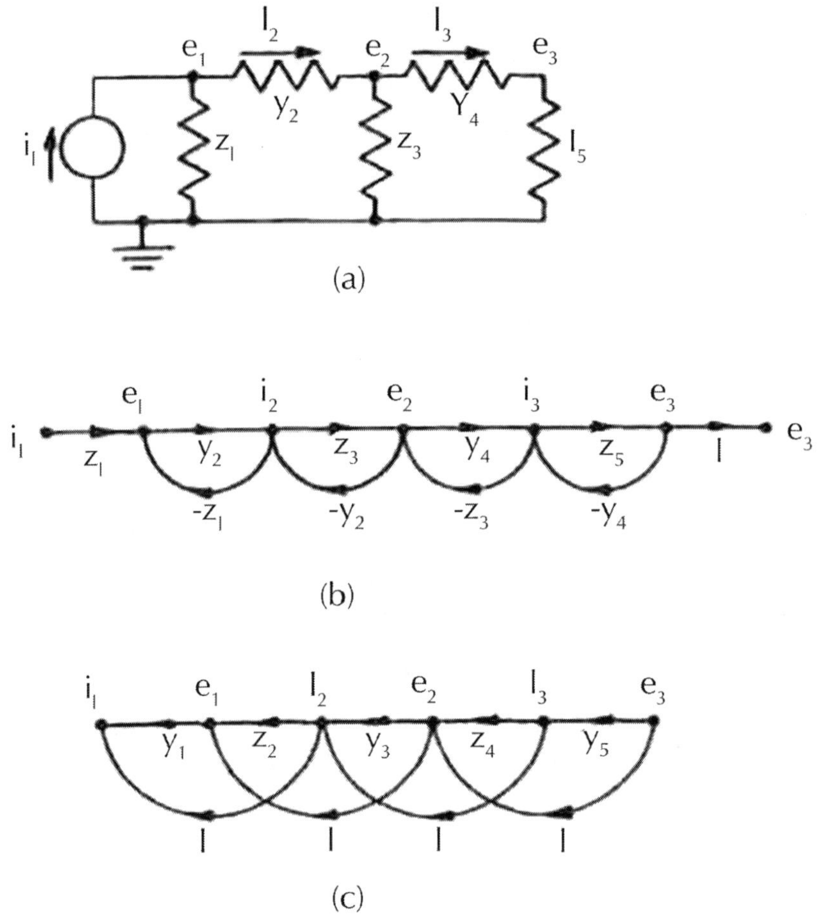

FIGURE 7 The transfer impedance of a ladder.

Consider the ladder network shownin Figure 7(a). The problemis to find the transfer impedance e_3/i_1. One possible formulation of the problem is indicated by the flow graph Figure 7(b).The associated equations state that $e_1=z_1(i_1 - i_2)$, $i_2=y_2(e_1 - e_2)$, and so forth. By inspection of the graph,

$$\frac{e_3}{i_1} = \frac{z_1y_2z_3y_4z_5}{1 + z_1y_2 + y_2z_3 + z_3y_4 + y_4z_5 + z_1y_2z_3y_4 + z_1y_2y_4z_5 + y_2z_3y_4z_5} \tag{13a}$$

or, with numerator and denominator multiplied by $y_1y_3y_5 = 1/z_1z_2z_5$.

$$\frac{e_3}{i_1} = \frac{y_2y_4}{y_1y_3y_5 + y_2y_3y_5 + y_1y_2y_5 + y_1y_4y_5 + y_1y_3y_4 + y_2y_4y_5 + y_2y_3y_4 + y_1y_2y_4} \tag{13b}$$

This result can be checked by the branch-combination method mentioned at the beginning of this chapter.

A different formulation of the problem is indicated by the graph of Figure 7(c), whose equations state that $i_3=y_5e_3$, $e_2 = e_3 + z_4i_3$, $i_2 = i_3 + y_3e_2$, and, so forth. In the physical problem $i_1$is the primary cause and e_3 the final effect.We may, however,*choose* a value of e_3 and then calculate the value of i_1*required* to produce that e_3. The resulting equations will, from the analysis viewpoint, treat $e_3$as a primary cause (source) and $i_1$as the final effect (sink) *produced* by the chain of calculations. *This does not in anyway alter the physical role of* i_1. The new graph (c) may appear simpler to solve than that of (b). Since, graph (c) contains no feedback loops, the determinant and path cofactors are all equal to unity. There are many forward paths, however, and careful inspectionis required to identify the sum of their gains as:

$$\frac{i_1}{e_3} = y_1z_2y_3z_4y_5 + y_1z_2y_3 + y_1z_2y_5 + y_1z_4y_5 + y_3z_4y_5 + y_1 + y_3 + y_5 \tag{13c}$$

Which proves to be, as it should, the reciprocal of Equation (13b). Incidentally, graph(c) is obtainable directly from graph (b), as areall other possible cause-and-effect formulations involving the same variables, by the process of path inversion discussed (Mason, 1953).This example points out the two very important facts: 1) The primary physical source does not necessarily appear as a source node in the graph, and 2) Of two possible flow graph formulations of a problem,the one having fewer feedback loops is not necessarily simpler to solve by inspection, since, it may also have a much-more complicated set of forward paths.

Figure 8(a) offers another sample analysis problem, determination of the voltage gainof a feedback amplifier. One possible chain of cause-and-effect reasoning, which leads from the circuit model, Figure 8(b), to the flow graph formulation, Figure 8(c), is the following. First notice that e_{01} is the difference of e_1 and e_k. Next express i_1 as an effect due to causes e_{oI} ande_k, using superposition to write the gains of the twobranches entering node i_1. The dependency of e_{0z} follows directly. Now, e_2 would be easy to evaluate interms of either e_{g2} or$i_f$if the other were zero, so, superpose the two effects as indicated by the two branches entering node e_z,. At this point in the formulation e_k and i_f are as yet not explicitly specified in terms of other variables. It is a simple matter, however, to visualize e_k as the superposition of the voltages in R_k caused by i_1 and i_f, and to identify i_f as the superposition of two currents in R_f caused by e_k and e_z. This completes the graph.

The path from e_k to e_{g1} to i_1 may be lumped in parallel with the branch entering i_1 from e_k. This simplification, convenient but not necessary, yields the graph showning Figure 9. We could, of course, have expressed it in terms of e_1 and e_k, at the outset and arrived at Figure 9 directly. All simplifications of a graph are themselves possible formulations.

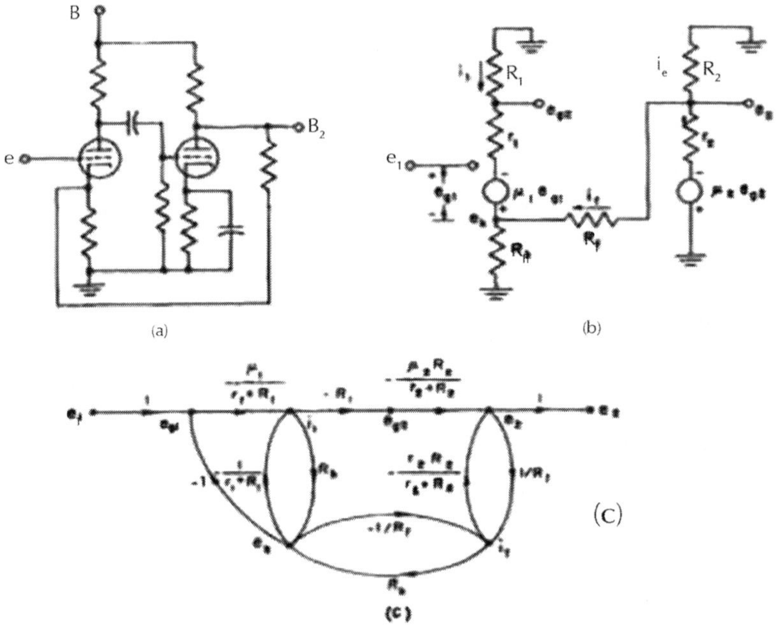

FIGURE 8 Voltage gain of a feedback amplifier. (a) A feedback amplifier, (b) The midband linear incremental circuit model, and (c) A possible flow graph.

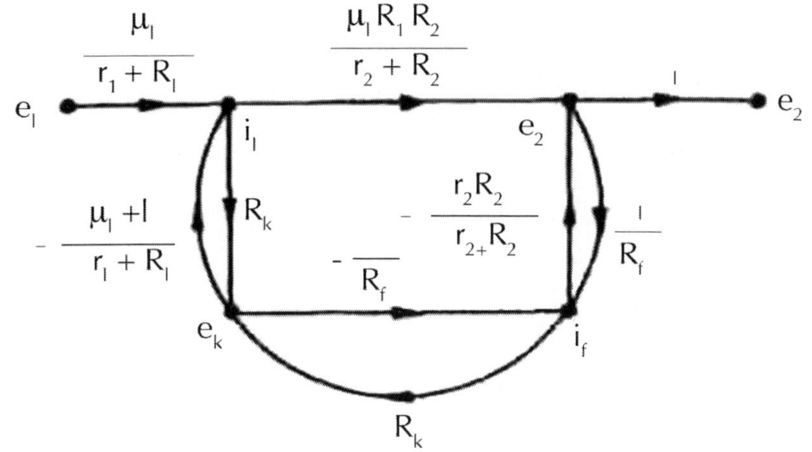

FIGURE 9 Elimination of superfluous nodes $e_{01}$and e_{02}.

The better perception of the workings of a circuit, the fewer variables will we need to introduce at the outset and the simpler will be the resulting flow graph structure.

In discussing the feedback amplifier of Figure 8(a) it is common practice to neglect the loading effect of the feedback resistor R_f inparallel with R_k, the loading effect of R_f in parallel with R_2, and the leakage transmission from e_k to e_2 through R_f. Such an approximation is equivalent to the removal of the branches from e_k to i_f and i_f to e_2 in Figure 9. It is sometimes dangerous to make early approximations, however, and in this case no appreciable labor is saved, since we can write the exact answer by inspection of Figure 9:

$$\frac{e_2}{e_1}=\frac{\dfrac{\mu_1\mu_2R_1R_2}{(r_1+R_1)(r_2+R_2)}\left[1+\dfrac{R_k}{R_f}\right]+\dfrac{\mu_1R_kr_2R_2}{(r_1+R_1)(R_f)(r_2+R_2)}}{1+\dfrac{(\mu_1+1)R_k}{r_1+R_1}+\dfrac{R_k}{R_f}+\dfrac{r_2R_2}{R_f(r_2+R_2)}+\dfrac{(\mu_1+1)\mu_2R_1R_kR_2}{R_f(r_1+R_1)(r_2+R_2)}+\dfrac{(\mu_1+1)R_kr_2R_2}{R_f(r_1+R_1)(r_2+R_2)}} \tag{14}$$

The two forward paths are $e_1\,i_1\,e_2$ and $e_1\,i_1\,e_k\,i_f\,e_2$, the first having a cofactor due to loop e_ki_f. The principal feedback loop is $i_1\,e_2\,i_f\,e_k$ and its gain is the fifth term of the denominator. Physical interpretations of the various paths and loops could be discussed but main purpose,to illustrate the formulation of a graph and the evaluation of its gainby inspection, has been covered.

As a final example, consider the calculation of microwave reflection from a triple-layered dielectric sandwich Figure 10(a) shows the incident wave A,the reflection B, and the four interfaces between adjacent regions of different material. Thefirst and fourth interfaces, of course,are those between air and solid. Let $r_1$be the reflection coefficient of the first interface, relating the incident and reflected components of tangential electricfield. It follows from the continuity of tangential *E* that the interface transmission coefficientis $1 + r_1$, and from symmetry that the reflection coefficient from the opposite side of the interface is the negative of r_1. A suitable flow graph is sketched in Figure 10(b). Node signals along the upper row are right going waves just to the left or right of each interface, those on the lower row are left-going waves, and quantities *d* are exponential phase shift factors accounting for the delay in traversing each layer.

Apart from the first branch r_1, the graph has the same structure as that of Figure 6(e). Hence, the reflectivity of the triple layer will be:

$$\frac{B}{A} = r_1 + (1 + r_1)(1 - r_1)G \tag{15}$$

Where *G* is in the same form as the gain of Figure 6(e). We shall not expand it in detail. The point is that the answer can be written by inspection of the paths and loops in the graph.

PROOF OF THE GENERAL GAIN EXPRESSION

A quantity Δ was defined as (Mason, 1953):

$$\Delta = (1 - T_1^{'})(1 - T_2^{'})\ldots(1 - T_n^{'}) \tag{16}$$

for a graph having *n* nodes, where,

$T_k^{'}$ = loop gain of the *k*th node as computed with all higher numbered nodes split.

Splitting a node divides that node into a new source and a new sink,all branches entering that node going with the new sink and all branches leaving that node going with the new source.The loop gain of anode was defined as the gain from the new source to the new sink, when that node is split.*It was also shown that Δ, as computed according to Equation (16), is independent of the order in which the nodes are numbered, and that consequently Δ is a linear function of each branch gain in the graph. It follows that is equal to unity plus the algebraic sumof various branch gain products.*

We shall first show that each term of Δ, other than the unity term, is a product of the gains of non-touching feedback loops.

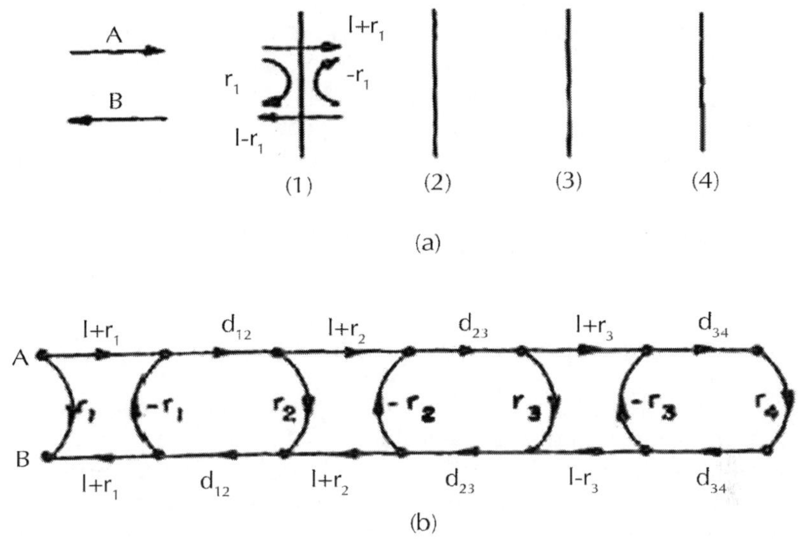

FIGURE 10 A wave reflection problem. (a) Reflection of waves from a triple-layer and (b) A possible flow graph.

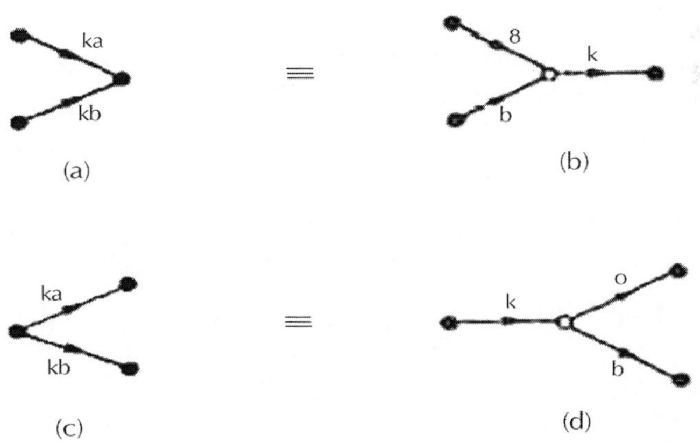

FIGURE 11 Two touching paths.

This can be done by contradiction. Consider two branches which either enter the same node or leave the same node, as shown in Figure 11(a) and (c). Imagine these branches imbedded in a larger graph, the remainder of which is not shown. Call the-branch gains *ka* and *kb*. Now, consider the equivalent replacements (b)and (d).The new node may be numbered zero, when $T_0^{'} = 0$, the other T' quantities in Equation

(16) are unchanged, and Δ is therefore unaltered. If both branches *ka* and *kb* appear in a term of the Δ of graph (a) then the square of *k* must appear in a term of the Δ of graph (b). This is impossible since must be a linear function of branch gain *k*. Hence, no term of Δ can contain the gains of two touching paths.

Now suppose that of the several non-touching paths appearing in a given term of Δ, some are feedback loops and some are open paths. Destruction of all other branches eliminates some terms from Δ but leaves the given term unchanged. It follows from Equation (16) and the definitions of $T_k^{'}$, however, that the Δ for the subgraph containing only these non-touching paths is just:

$$\Delta = (1 - T_1)(1 - T_2) \ldots (1 - T_m) \tag{17}$$

Where T_k is the gain of the *k*th feedback loop in the subgraph. Hence, the openpath-gains cannot appear in the given term and it follows that each term of is the product of gains of non-touching feedback loops. Moreover, it is clear from the structure of Δ that a term in any subgraph Δ must also appear as a term in the Δ of the complete graph, and conversely, every term of Δ is a term of some subgraph Δ. Hence, to identify all possible terms in Δ we must look for all possible subgraphs comprising sets of non-touching loops. Equation (17) also shows that the algebraic sign of a term is plus or minus in accordance with an even or odd number of loops in that term. This verifies the form of Δ as given in Equation (12c) and Equation (12d).

We shall next establish the general expression for graph gain Equation (12a). The following notation will prove convenient. Consider the graph shown schematically in Figure 12, with node *n*+ 1 given special attention.

Let:

Δ' = the A for the complete graph of n + 1 nodes.

Δ = the valueof Δwithnode $n + 1$ split or removed.

T = the loop gain of node $n + 1$.

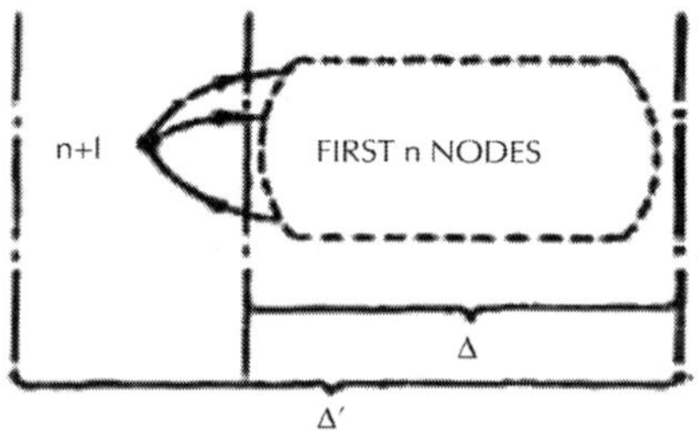

FIGURE 12 A flow graph with one node placed strongly in evidence.

There will in general be several different feedback loops containing node $n + 1$. Let:

T_k = gain of the kth feedback loop containing node $n + 1$

Δ_k = the value of Δ for that part of the graph not touching loop T_k
with the notation, we have form Equation (16) that L:

$$1 - T = \frac{\Delta'}{\Delta}. \tag{18}$$

Remembering, that any Δ is the algebraic sum of gain products of non-touching loops, we find it possible to write.

$$\Delta' = \Delta - \sum_b T_k \Delta_k \tag{19}$$

Equation (19) represents the count of all possible non-touching loop sets in Δ'. The addition of node n + 1 creates new loops T_k but the only new loop sets of Δ' not already in Δ are the non-touching sets $T_k\Delta_k$. The negative sign in Equation (19) suffices to preserve the sign rule, since, the product of T_k, and a positive term of Δ_k will contain an odd number of loops.

Substitution of Equation (19) into Equation (18) yields the general result:

$$T = \frac{\sum_k T_k \Delta_k}{\Delta}. \tag{20}$$

With node n + 1 permanently split, T is just the source to-sink gain of the graph and T_k is the kth forward path. This verifies Equation (12a).

CHAPTER 19

AN OVERVIEW OF SIGNAL FLOW GRAPHS

KHOMAN PHANG

AN OVERVIEW OF SIGNAL-FLOW GRAPHS

The signal-flow graphs have long been used in many areas of engineering. Originally devised by Mason for linear networks (Mason, 1953), they are a mainstay of network theory and are commonly applied to areas as diverse as automatic control and data communications. This section provides an overview of linear signal-flow graphs, largely for the benefit of today's reader who may not have had much exposure to network and graph theory. Much of the following material is derived from (Haykin, 1970) and the reader is also referred to (Mason, 1960) and (Chen, 1991 and1997) for a more thorough treatment of this fascinating area.

A graph is a collection of points and lines, respectively referred to as *nodes* and *branches*. Each end of a branch is connected to a node and both ends of a branch may be connected to the same node. A signal-flow graph is a diagram which depicts the cause and effect relationship among a number of variables. The variables are represented by the nodes of the graph, while the connecting branches define the relationship.A typical signal-flow graph is shown in Figure 1. The figure has four nodes, each representing a node signal x_j. Between a pair of nodes j and k lies a branch with a quantity called the branch transmittance t_{jk}, represented here by theletters a to f.

From Khoman Phang, CMOS Optical Preamplifier Design Using Graphical Circuit Analysis, PhD Thesis, Dept. of Electrical and Computer Engineering, University of Toronto, June 2001. Used with permission.

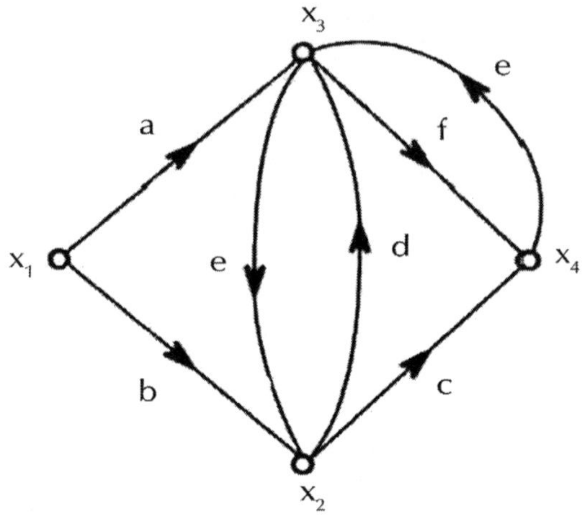

FIGURE 1 A linear signal-flow graph.

The flow of signals in the various parts of the graph is dictated by the following three basic rules which are illustrated in Figure 2:

1. In Figure 2, a signal flows along a branch only in the direction defined by the arrow and is multiplied by the transmittance of that branch.
2. Figure 2, a node signal is equal to the algebraic sum of all signals entering the pertinent node *via* the incoming branches.
3. Figure 2, the signal at a node is applied to each outgoing branch which originates from that node.

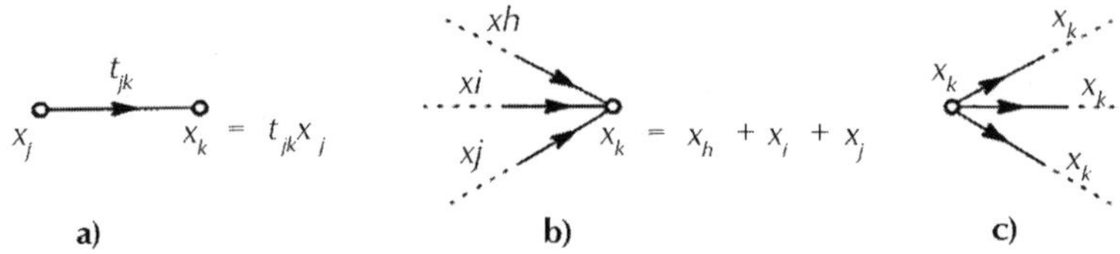

FIGURE 2 Illustrating three basic properties of signal-flow graphs.

From these basic rules are derived the four elementary equivalences shown in Figure 3, which guide one in the manipulation of signal-flow graphs. These equivalences are sufficient for the complete reduction of a graph containing no feedback loops. To handle graphs that incorporate feedback loops, there are two additional equivalence relations. Consider a self-loop in which a node signal is fed back to itself as illustrated on the left-hand side of Figure 3.

The signal-flow graph represents the relation:

$$x_3 = x_2 = Lx_2 + x_1 \tag{1}$$

from which x_3 can be expressed exclusively in terms of x_1 as:

$$x_3 = \frac{1}{1 - L} x_1 \tag{2}$$

And represented by the right-hand side of Figure 3 (a). Figure 3 (b) illustrates the classic feedback structure comprised of a gain stage A, surrounded by a feedback network β. The signal-flow graph represents the pair of equations:

$$x_3 = Ax_2 \tag{3}$$

$$x_2 = \beta x_3 + kx_1 \tag{4}$$

from which we obtain the familiar expression:

$$\frac{x_3}{x_1} = \frac{kA}{1 - A\beta} \tag{5}$$

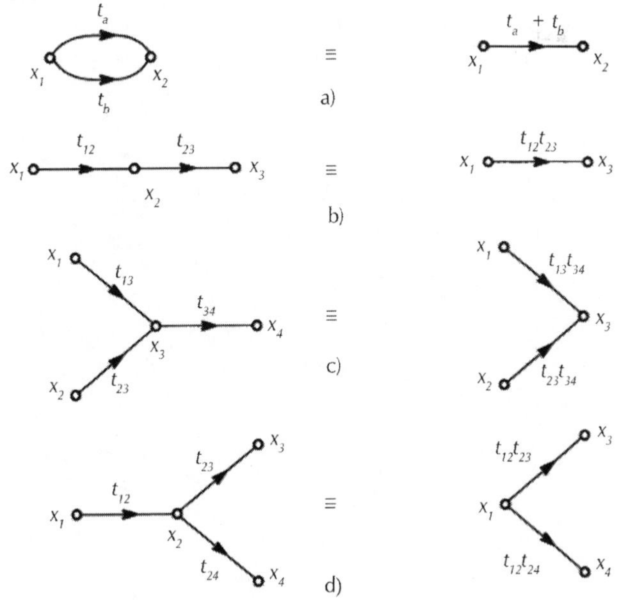

FIGURE 3 Four elementary equivalences of signal-flow graphs.

In contrast to Equation (2) and Equation (5), most textbooks have a plus rather than minus sign, a result of adopting a convention where by the feedback signal is subtracted rather than added back to the input node. Since, the difference is only one of convention, we will continue with existing convention in order to remain consistent withthe signal-flow graph algebra.The quantities L and $A\beta$ are commonly known as the loop gain.

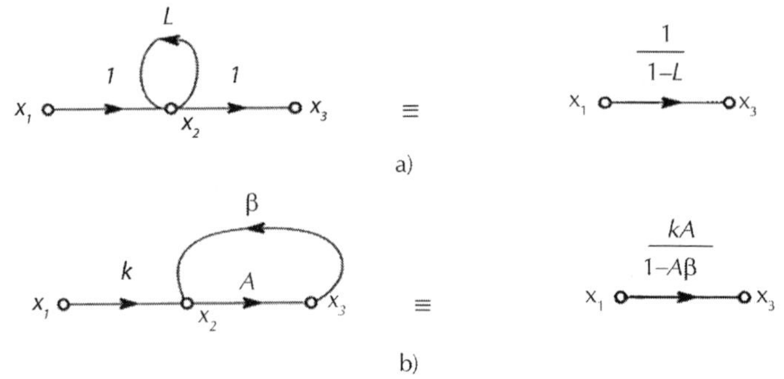

FIGURE 4 Collapsing feedback loops:(a) self-loop and (b) general feedback loop.

By using the elementary equivalences in Figure 3 and Figure 4,any transfer function can be derived from a signal-flow graph by successively collapsing internal nodes until only the input and output nodes remain. Figure 5 illustrates this process. The resulting transfer function is:

$$\frac{x_{out}}{x_{in}} = \frac{abc}{1 - cd - bce}. \tag{6}$$

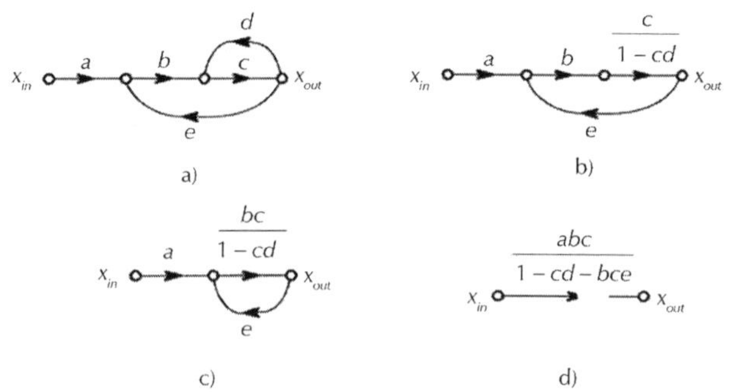

FIGURE 5 Determining a transfer function through collapsing of signal flowgraph.

MASON'S DIRECT RULE

The manipulation of signal-flow graphs is an effective and straight forward means of determining transfer functions for relatively small graphs. However, such manipulations quickly become unwieldy for larger graphs, and for such situations the transfer function can be computed directly. Comparing Equation (6) to the original signal-flow graph in Figure 5 (a), we notice that the transfer function can be expressed as:

$$\frac{x_{out}}{x_{in}} = \frac{P_1}{1-(L_1+L_2)} \tag{7}$$

Where $P_1 = abc$ representsthe forward transmission path from input to output, and $L_1 = cd$ and $L_2 = bce$ represent the loop gains of the two feedback loops found in the graph. In general, the transfer function of a signal-flow graph can bederived using the following expression, commonly known as Mason's Direct Rule (Mason, 1960):

$$\frac{x_{out}}{x_{in}} = \frac{1}{\Delta}\sum_{k=1}^{n} P_k\Delta_k \tag{8}$$

Where:

P_k = Transmittance of the *k*th forward path from input x_{in}, to output, x_{out}

Δ = 1 – (sum of all individual loop gains) + (sum of loop gain products of all possible sets of non-touching loops taken two at a time) – (sum of loop gain products of all possible sets of non-touching loops taken three at a time) +........

and

Δ_k = the value of Δ for that portion of the graph not touching the *k*thforward path.

SUMMARY

In this chapter, we discussed the photodetector and optical preamplifier that make up the front-end of an optical receiver. The trans impedance amplifier is the most common preamplifier structure, and we described the three principal new requirements that we wish to address: A wide dynamic range, ambient light rejection, and low-voltageoperation. In preparation for discussion of a graphical circuit analysis technique, we reviewed existing analysis techniques such as nodal analysis, and feedback analysis based on amplifier topology and return ratios. In addition, we reviewed the basic conventions of signal-flow graphs and outlined Mason's Direct Rule.

CHAPTER 20

A THEORY ON NEUROLOGICAL SYSTEMS-PART I AND PART II

BRADLEY S. TICE

INTRODUCTION

In re-addressing points made by Penrose about artificial intelligence in his book *The Emperor's New Mind* (1989) a new view arises that gives a formal structure to a new model of neurological systems. It deals with questions of artificial intelligence. Hoped to narrow some of these arguments and from that give substance to a new model of neurological systems. It will focus on three specific areas: 1. The Turing model of intelligence in computers, 2. Godel's Theorem and numbers, and 3. Organic and non-organic systems.

PART I

In Alan Turing's paper "Computing Machinery and Intelligence" (1950) he considers the "imitation game" as a test for whether a computer can pass for a human being as being a valid standard for a machine being intelligent Penrose finds the Turing test of machine intelligence a "valid indication of the presence of thought, intelligence, understanding, or consciousness is actually quite a strong one" (Penrose, 1989) and "Thus I am, as a whole, prepared to accept the Turing test as a roughly valid one in its chosen context" (Penrose, 1989). Now, while Penrose is clear to point out the weaknesses found in the Turing test of machine intelligence, such that a computer must be able to imitate a human but a human does not have to imitate a computer (Penrose, 1989), he has still accepted it as a test of a machine's ability to have the following human properties—thought, intelligence, understanding, and consciousness.

In my paper "The Turing Machine: A Question of Linguistics?" I raise the issue that the Turing test of intelligence is just a question of language rather than intellect and even such factors as the type of language used, the physical and cultural abilities and knowledge of a human may cause a person to "fail" this type of intelligence test (Tice, 1997/2004) Penrose has placed too many extraneous factors on this type of test and seems to miss the point that real intelligence is something beyond a game.

Mason, S.J. (1955) Feedback Theory – Further Properties of Signal Flow Graphs. Technical Report 303, MIT. Reprint from Proceedings of the I.R.E. Vol. 44, No. 7, July 1956. Reprinted with permission from the Institute of Electrical and Electronics Engineers.

PART II

Penrose uses Godel's Theorem as a "proof that mathematical insight is by nature non-algorithmic(Penrose, 1989). Unfortunately, Penrose has confused the fact that because Godel's Theorem states that not all axiomatic propositions can be proved, and hence, the thought process used for such thinking is non-algorithmic, the nature of mathematical "insight", the action of realizing that a non-algorithmic process is itself a non-algorithmic process, gives light to the reason to suppose that human thought is nonalgorithmic (Penrose, 1989).

In my book *Formal Constraints to Formal Languages* (In Press), I address the question of Godel's Theorem and Hilbert's axiomatic foundations and that it did not provide an "absolute" factor to the provability of propositions of number theory (Tice, In Press). Also the use of a Universal Truth Machine (UTM) is given to present the basic procedure for Godel's Incompleteness Theorem (Tice, In Press). An interesting result occurs when I substitute the words 'will never' with the word "may" in the following sentence from theUTM:

UTM will never say G is true.

Resulting in the following sentence:

UTM may say G is true.

What results is that the UTM becomes universal and changes the primary strengths found in Godel's Incompleteness Theorem, namely that some propositions can be axiomatically proved and some may not but the robustness of the axiomatic system stays intact because it has accounted for such variants.

PART III

In von Neumann's paper "Probabilistic logics and the synthesis of reliable organisms from unreliable components" (1956/1963).The outlines the logical structure of reliable systems. In this paper he points out that the concept of "error" must be viewed as not an extraneous and misdirected "accident" but as an essential part of the process which is normal in the correct logical structure of such a system (Taub, 1963). Von Neumann then proceeds to design automata that parallel a synthetic system, relays and circuits, with organic, neurons and the nervous system (Taub, 1963).

An interesting point is if one considers numbers, all types, to be "components" of a system and those that are functionally problematic, such as algorithms that "maybe" provable but are, for some reason, intractable, become "errors" within the system and hence, can be a unreliable part that can be replaced. In a system that is robust, numbers and algorithms that are "problematic" are replaced by functional ones and allow the system to keepoperating.

SUMMARY

In summarizing, the following can be noted; that the Turing test for intelligence in machines is not viable, that Godel's Incompleteness Theorem is not an impediment to mechanical algorithmic functions if there is a selection of functional ones to replace the one's that are intractable. Organic systems, natural, may provide a systematic method for non-organic systems, mechanical, to follow in the function, rather than design, of a man made system.

PART II

INTRODUCTION

Application of traditionalmechanistic approaches found in the fields of artificial intelligence and engineering can be used to develop systematic models of neurological systems. In a transitional development from an earlierpaper, the mechanistic notions found in artificial intelligence can be used to help elaborate a systematic approach to the study of neurologicalsystems (Tice, 2005). The resulting effect will be a more structured development of the notions found in neurological processes and will complement concepts of systems found in the engineering disciplines.

PART I

The growing field of neurological systems is a development from systems biology that has its roots in the physicalsciences and engineering fields. The nature of a system is the systematic description of the process or processes being defined as found in an organized hierarchy of those processes. Current studies in neuroscience have moved beyond the notion of the neuron acting as a single functional unit to that of complex systems with many variables (Bullock, Bennett, Johnston,Josephson,Marder, and Fields, 2005). A systems level approach has been proposed as the next stage of study to bridge thegap between molecular units and psychology in the neural sciences (Stem and Hines, 2005).

PART II

With this new focus on systems level processes the need to understand those methods used to understand systems evaluation will fall into the areas of engineering. As artificial intelligence has a long history of trying to "mirror" the naturalprocesses of thought and has a strong back ground as an interdisciplinary outgrowth of electricalengineering, it would seem reasonable to assume that some applications may be applied to the study of neuroscience. Even a systems approach to neuroscience has its limitations as was found in the study of artificialintelligence in the 1960's when simple systems applications, servo-mechanisms and analog networks, could no longer be developed to the next stage of inquiry (de Callatay,1992). Only with the advent of algorithmic languages and computer simulations could the current growth in neural networks research be done (de Cattalay, 1992). Even the use of computers to generate large amounts of data has the inherent problems of analyzing such copious amounts of data by researchers. The need for new methods for analysis will only grow.

SUMMARY

It is clear that the application of a systematic type of modeling of processes found in neurologicalsystems parallels many concepts found in artificial intelligence and is made more pronounced, and profound, in that such developments fore-shadow a more common ground between nature and mechanical analysis of b oth biological, neurological, and engineered, artificial intelligence systems.

CHAPTER 21

A THEORETICAL MODEL OF FEEDBACK IN PHARMACOLOGY USING SIGNAL FLOW DIAGRAMS

BRADLEY S. TICE

INTRODUCTION

The dissertation will model a theoretical system of pharmacological drug feedback by the use of signal flow diagrams. A comprehensive evaluation of a signal flow diagram and then a comparison with traditional formal or block diagrams will be examined. A clinical study using specific drugs and specific problems, immunological studies taken from actual clinical case studies, will be used to model the feedback system with the signal flow diagrams.

They will be in three separate segments, Model A, B, and C of which Model A will discuss delayed hypersensitivity skin testing. Model B will be the triple test plan for serologic diagnosis of syphilis and Model C the laboratory effects of cyclosporine. The results from this study will present the signal flow diagram as a clear, accurate, and easy method of pharmacological feedback.

A practical application can be made for using signal flow diagrams to graph pharmacology feedback by the health professions along the lines of medical flow chartsthat are used to reduce errors, burnout, and redundancy in patient care (Cronin Lane and Peirce, 1983).

LITERATURE SURVEY

Feedback as defined by the engineering concepts of control systems started in the 1920'sand 1930's and was first given voice in Minorsky'spaper on the steering of ships in 1922 and followed by Nyquist's paper on Regeneration Theory, 1932, and Hazen's paper on the Theory of Servomechanisms in 1934 (Chang, 1961). Brown and Whiteley's work on servo mechanism theory and Tustin'swork on non-linear elements in servo-design are also worth mentioning (Porter, 1950). The first practical application of feedback control was James Watt's flyball governor to the steam engine in 1775 (Murrill, 1967).

A system is defined as an orderly combination or arrangement of parts into a whole and that these combination of parts represent a methodological arrangement that represent a system (Koenig and Blackwell, 1961). A feedback control system is a combination of elements which cooperate to maintain a physical quantity, the output, equal to the desired output that is related to other physical quantities known as inputs (Newton and Gould, 1957). A feedback system is distinguished from a network by the presence of at least one unilateral element that represents power, information, or a commodity that can flow in only one direction (Smith, 1958).

All automatic regulating systems can be divided into two groups, the direct-acting and the indirect-acting systems. The direct-acting systems is where the action of the sensitive element on the regulatory organ is operated with the introduction of additional source of energy and the indirect system is when the sensitive element acts on the regulatory organ directly through a special amplifier as an auxiliary source of energy (Popov, 1962). Feedback can be considered a form of communication in which an input is responded to by an output (James, Nichols, and Phillips, 1964).The loss of time between actions and reactions is the major point of control systems problems and represents the measure of adisturbance (Holzbock,1958).

A closed-cycle control system is any signal that is supplied to the controller as a function of the object or device being controlled (Zeines,1959), (Horowitz,1963), and (Ku, 1962). A physical system is stable if, when disturbed from its equilibrium state, it can ultimately return to that original condition, it is unstable if it increases indefinitely with time (Hardie, 1964). The parameters of such a system must be "optimally adjusted" to insure optimum control action and is the heart of the feedback system (Oldenbourg and Sartorius, 1953) and (Kipiniak, 1961).

The selection of these primary measuring elements is important in that they must conform to the control requirements of the process in controlling the variables (Hadley and Longobardo, 1963). Most control feedback systems are defined as servo mechanisms and represent a closed-loop system (Chubb, 1967) and (Gille, Pelegrin, and Decauline, 1959). The elements which constitute the essentials of a servo mechanism are also the inherent properties of a vibrating system and can be detrimental to a servo mechanism (Bulliet, 1967). A power control system is considered to be an object of automation and servo-systems the tools to automation (Bulgakov, 1965).

The sole purpose of a control system is to minimize the "process-upset" when such a disturbance arises (Tucker and Willis, 1958). The use of "setpoints" as an ideal or specified process are the guidelines used to keep the control in a system and are used to measure all deviant actions to maintain control (Rusinoff, 1957). Modeling of such a system allows for the evaluation of systems that would otherwise be beyond the direct observation of a feedback system but special care must be taken to insure that the model accurately describes the function it is modeling (Seifert and Steeg, 1960) (Derusso, Roy and Close,1965).

Feedback systems are all dynamic systems and fundamentally similar and operate on the difference between the actual state of the system and the arbitrarily varied ideal state (Ahrendt, 1954) and (Bowen and Schultheises, 1961). Feedback is also a form of self-regulationand are an inherent process in both living and non-living processes of nature (Nagel, 1948). In defining sub-units of a feedback system, elementary components should be defined by their functions rather than by appearance and that before a physical quantity can be controlled, it is essential that it be measurable (Warren,1967), (West, 1953), and (Qvamstrom, Schurt, and Runnstrom-Reio, 1965).

The mathematical measurement of feedback systems is usually the domain of differential equations in that such processes are based on a present state that is instantaneous rather than a past history of actions (Oguztoreli, 1966). The use of linear differential equations assumes the linearity of the properties being formulated.Although, such characteristics do not reflect the true nature of the real world as no system is completely linear or non-linear also the mathematical modeling of systems is difficult when working with "optimizing" and "adaptive" systems (Thaler and Pastel, 1962) and (Peschon, 1965).

A differential equation is linear if the equation contains only first powers of the dependent variable or its time derivatives (Tsien, 1954). A non-linear equation is the result of higher powers of the variable and cross products of these variables and its derivatives (Tsien, 1954). In designing an "optimal" control system, four kinds of variables must be accounted for in the design. Independent variables which are the manipulating variables for the control or monitoring of the process. Dependent variables that serve to measure and describe the state of the process at any instant of time. Product variables that are used to indicate and measure the quality of the operating performance of the control system and fourth, the disturbances which are the uncontrollable, environmental variables (Tou, 1964).

The use of the Root Locus technique is used to determine time domain properties and is a graphical method of determining the roots of the characteristic equation of a single loop system (Kuo, 1962). The use of Fourier and Laplace transformations are used to analyze transient processes in linear systems where signals appear as prescribed functions of time (Solodovnikov,1965). The two most important determining factors of the use of differential equations is what is the type of system and the manner in which the specifications are set for has performance of the system can only be examined by the "appropriate" solutions to these equations (Wilts, 1960).

The mathematical definition of adaptability is not a precise one due to the fact that feedback has various definitions and permutations (Mishkin and Braun, 1961). Physical systems have many common characteristics that can be described by mathematical modeling and assigned definite groups according to the structure of these models (Tomovic, 1963). Feedback adds to the complexity of a system and hence adds to the complexity and difficulty of the analysis of such systems (Lindorff, 1965).

Another term for feedback systems is cybernetics. The leading figure in the field of cybernetics is the mathematician Norbert Wiener and is best embodied in his famous book *Cybernetics* (Cambridge: MIT Press, 1948). Wiener states "Messages are themselves a form of pattern and organization". Indeed, it is possible to treat sets of messages as having entropy like sets of states of the external world. Just as entropy is a measure of disorganization, the information carried by a set of messages is a measure of organization. In fact, it is possible to interpret the information carried by a message as essentially the negative of its entropy, and the negative logarithm of its probability (Wiener, 1950).

Another term for feedback is homeostasis. Cybernetics is a Greek word for pilot or steers man. Homeostasis is the balancing of the bodies systems to produce and support life. Strict limits, especially when dealing with higher life forms, is necessary in defining the conditions that make up the properties of homeostasis and tend to operate more slowly than the voluntary or postural feedbacks (Wiener, 1948). William Cannon's seminal work, the wisdom of the body, is a prime example of homeostasis in the human body and cites clear examples of these processes operating in the organsand systems of man (Cannon, 1932).

The nervous system is a system which controls the interrelationship between the organism and the environment and regulates the internal environment, the process of homeostasis (Adolph, 1960). Although, Cannon was the first to use the term homeostasis, he paid homage to the German physiologist Pfluger and the Belgian physiologist Fredericq and the ancient Hippocrates for the generation of this concept (Langley, 1965).

Cannon considered homeostasis as a defense protection of the body by the preservation of consist encyin the fluid matrixofthebodyagainstunfavorable conditionsthat would arise if not preserved (Cannon, 1932). Homeostasisis derived from the term "homeo" which means like or similar, and "stasis" meaning standing. Metabolism has also been applied to the "inflow" and "outflow" of matter and energy and intermediary transformations within the organism (Henderson, 1913).

Another definition of feedback is communication theory and information science. Lwoff has defined informationas following Brillouin's analysis, if considering a certain number of possible answers, when no informationis available, and when some information is gained, then the number of possible answers is reduced, and the complete information means only one answer, in that information is a function of the ratio of possible answers "before and after" (Lwoff, 1962). Communication theory is best represented by Shannon and Weavers seminal work.The mathematical theory of communication (Urbana: University of Illinois Press, 1949) that like Norbert Wiener's Cybernetics, helped define early communication theory.

Information science is a general term used to cover thefields of library science, computer networks and cognitive and neural sciences. Types of feedback are represented in all of these disciplines. Two works that deal with information, as a quantitative science are Edward R. Tufte's, visual display of quantitative information and envisioning information (Tufte, 1983 and 1990).

The basic characteristics of a feedback system, as defined as an automatic-control system, can be illustrated qualitatively by the use of a block diagram (Thaler, 1955). A block diagram indicates a signal transfer where a circuit diagram represents also the state of energy transfer (lzawa, 1963). Feedback control systems function in the human body and represent some of the most complexsystems in the world (Instrument Society of America, 1958).

All types of feedback control systems can be described in terms of functional block diagrams.The blocks in a diagram are interpreted as representing functions of components and not as isolated pieces of equipment as several functions can be combined into a single piece of physical equipment or several pieces will be required to perform a single function (Doebelin, 1962). The block diagram represents the operations performed in a system and in a manner in which the signal information flows throughout the system (DelToro and Parker, 1960). A block diagram is a graphical representation of inter connected elements or components which form a system and differ only in their dynamic properties (Solodov and Fuller, 1966).

The signal flow diagram has the conventional form of a circuit diagram, comprising connected branches and nodes, and differs from the circuit diagram in two important ways. The nodes are the points where the signals appear and where they are summated, and that the branches are oriented and their orientation forms a single channel for the travel of the signal (Macmillian, 1964).

The Figure 1 is a graphical representation, block diagrams, of a basic feedback system (West, 1960).

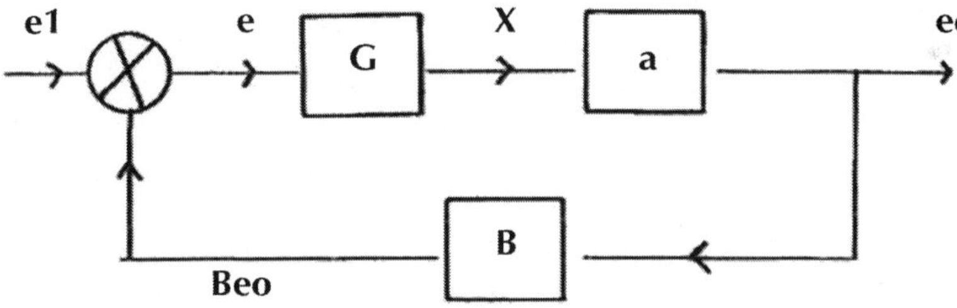

FIGURE 1 Graphical representation of a basic feedback system.

The original amplifier with gain ‘a’ has an input and output signal ‘x’ and an output e related by eo = a x.An additional amplifier of gain ‘G’has an input and output signals ‘e’ and ‘x’ respectively with x = Ge. The signal e is feedback through a passive network B to form a feedback signal Beo. A comparator or error detector indicated by a circle is fed with an input signal e and the feedback signal Be to form the signal ‘e’ defined by e = ei–Beo.The whole system is expected to operate as an amplifier of gain ‘a’ with improved reliability (West, 1960).

This is an example of a generic diagram of a adaptive control system.

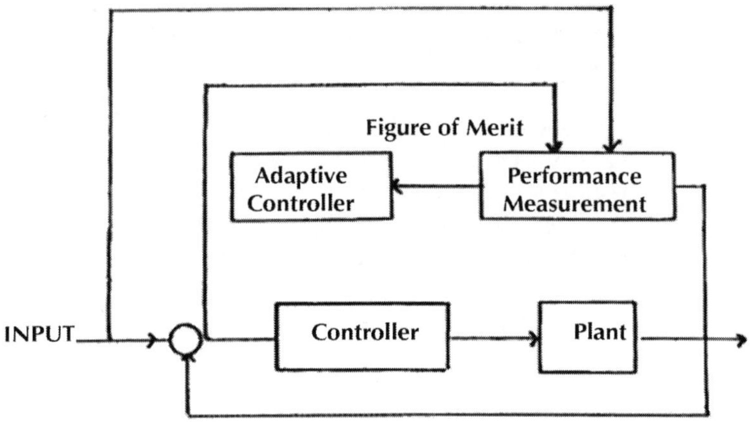

FIGURE 2 Generic diagram of adaptive control system.

The adaptive control system has three necessary performance measurements (Caruthers and Levenstein, 1963).

1. Means for a performance measurement.
2. Means for translatingthisintoaquantitative figure of merit.
3. A closed loop control of system parameters to achieve an optimum as given by the figure of merit.

The performance measurement is assumed to be based on observation of system input, output, and/or error signal.The figure of merit or error signal is derived from these through the definition of the performance criterion. The adaptive controller adjusts controller parameters to achieve an optimum (Caruthers and Levenstein, 1963).

These two diagrams are to give a visual reference to feedback and adaptive control systems.

In reviewing biomedical papers using feedback systems the following was found. The use of Bayesian networks for drug delivery optimization (Bellazzi, 1993). Biological modeling on a microcomputer using a standard spreadsheet and equation solver programs (Plouffe, 1992). Advanced computer programs for drug dosing that combine pharmacokinetic and symbolic modeling of patients (Lenert, Lurie, Sheiner, Coleman, Klostermann, andBlaschke, 1992).

Open-loop stochastic control of pharmacokinetic systems (Lago, 1992). Analysis of low-frequency lung impedance in rabbits with non-linear models (Peslin, Saunier, Duvivier, and Marchand, 1995). An alternative pathway for signalflowfromrodphotoreceptorsto ganglion cells in mammalian retina (DeVries and Baylor, 1995). Validation of continuous thermal measurement of cerebral blood flow by arterial pressure change (Wei, Shea, Saidel, and Jones, 1993). Problems of selecting the reagent concentration in the solid-phase immune enzyme method for determining antibody concentrations using block diagrams (Panteleev, Vaneeva, Demchenko, and Semenova,1987).

A new design concept for an audio dosimeter described by block diagrams (Clark, 1980). Conceptions for optimization in radiation therapy using block diagrams (Hubener, 1978). Contrast echocardiography in clinical practice (Dubourg, Chikli, and Delorme, 1995). A new analysis method for disposition kinetics of enterohepatic-circulation of diclofenac in rats (Fukuyama, Yamaoka, Ohata, and Nakagawa, 1994). Discussion of the charging and discharging circuits and present the general block diagram of an automatically triggered current defibrillator (Monzon and Guillen, 1985). Differences between normal blood pressure regulation in the circulatory system under normal, orequifinal conditions areanalyzed using hydraulic models, block diagrams, and signal flow diagrams (Guenther, Morgado, and Penna, 1974).

Testing fiberoptic data links for sensitivity to high gamma radiation dose rates using block diagrams (Krinsky, 1988). A simple device for photographic dosimeter-calibration using block diagrams (Laduand Randaccio, 1980). A more simple PROM programmer using block diagrams (Coco, 1979). Measuring, controlling, and dose-rating with miniature oval-gear counters using block diagrams (Feil, 1979). New developments in X-ray television using block diagrams (Heister, 1979). Modeling, testing, simulation, and optimization of digitalis pharmacokinetics using block diagrams (Ranjbaran, 1974). Models of functional shutdown of 1 hemisphere and neuropharmacological effects on deep cerebral structures using block diagrams (Menshutkin, Suvorova, and Balonov, 1981).

Television microscopes for studying biological micro scopicmaterial using block diagrams (Danilov, 1984). State variable techniques in pharmacokinetics using computer graphics and block diagrams (Ranjbaran, 1980). The papers dealing with graphical information of feedback almost exclusively use block diagrams for visual representation and this follows the long tradition of using such methods in the fields of engineering, information science, and design.

SIGNAL FLOW DIAGRAM

The use of signal flow diagrams are common in fields such as engineering and a practical use of themcanbe made in the field of pharmacology. The main reason for the use of signal flow diagrams over other diagram systems, formal or block diagrams, are that they are easy to use and permits a solution practically upon visual in spection (Shinners, 1964)[1].

Signal flow diagrams can solve complex linear, multiloop systems in less time than either block diagrams or equations (Macmillian, Higgins, and Naslin, 1964). A signal flow graph is a topological representation of a set of linear equations as represented by the following equation:

Equation 1: Yi = L aij Xj, i= 1............n

Branches and nodes are used to represent a set of equations in a signal flow graph. Each node represents a variable in the system, like node i represents variable y in equation 1.Branches represent the different variables such as branch ij relates variable yi to yj where the branch originates at node i and terminates at node j in equation 1 (Shinners, 1964).

The following set of linear equations is represented in the signal flow graph in Figure 3 (Shinners, 1964).

$y_2 = ay, + by2 + by3$
$y_3 = dy2$
$y_4 = ey1 + fy3$
$y_5 = + gy3 + hy4$

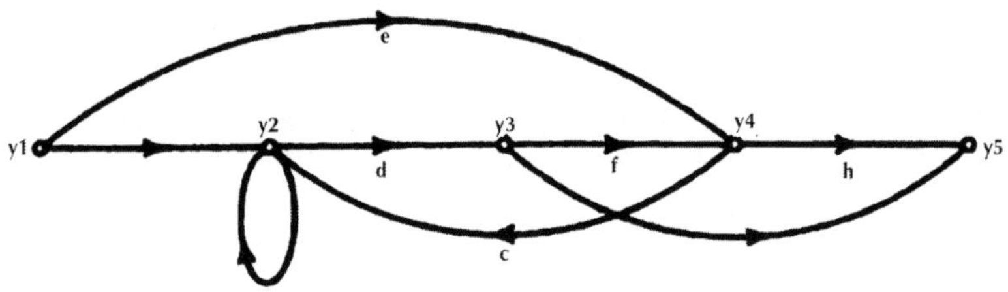

FIGURE 3 Signal flow graph.

It is necessary now to define the terms as represented by the signal flow diagram in Figure 3 (Shinners, 1964).

1. The source is a node having only outgoing branches, as y_1 in Figure 3.
2. The sink is a node having only incoming branches; as y_5 in Figure 3.
3. The path is a group of connected branches having thesame sense of direction. These are he, adfh, and b inFigure 3.
4. The forward paths are paths which originate from a source and terminate at a sink along which no node is encountered more than once, as are eh, adg, and adfh in Figure 3.
5. The path gain is the product of the coefficient associated with the branches along the path.
6. The feedback loop is a path originating from a node and terminating at the same node. In addition, a node cannot be encountered more than once. They are b and dfc in Figure 3.
7. The loop gainis the product of the coefficients associated with the branches forming a feedback loop.
8. By using a signal flow diagram to represent the variables associated with pharmacological testing, drug delivery, behavior, dosage and time intervals can all be graphed for ease of representation of these complex systems.

SIGNAL FLOW VERSES BLOCK DIAGRAMS

The strongest points to be made in favor of the signal flow diagram over block diagrams, sometimes also termed formal diagrams, is that they are more general in use, being applied in a more collective or universal manner, and that they are a more pronounced rationalization stemming from diagrammatics implification (Macmillian, Higgins, and Naslin, 1964).

A block or formal diagram is a graphical representation of interconnected elements which form a system that differ only in their dynamic properties (Solodov, 1966). A signal flow diagram represents a set of equations by means of branches and nodes. The following are examples of how complex and visually daunting block diagrams can become with multiple equations (Shinners, 1964).

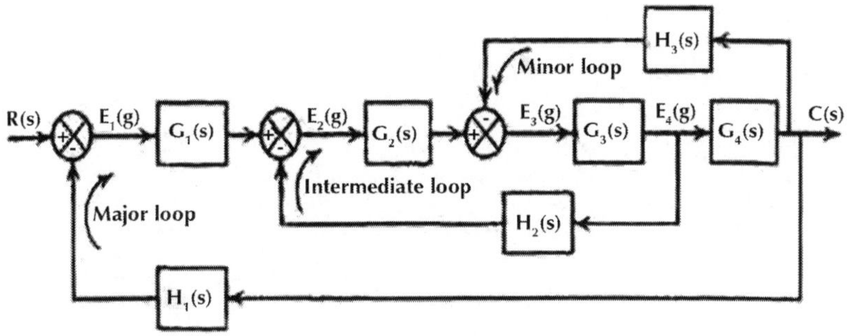

FIGURE 4 Original system.

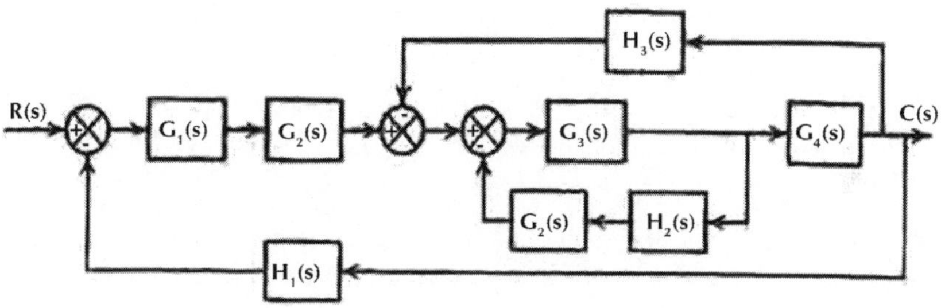

FIGURE 5 Rearrange the summing points of the intermediate and minor loops.

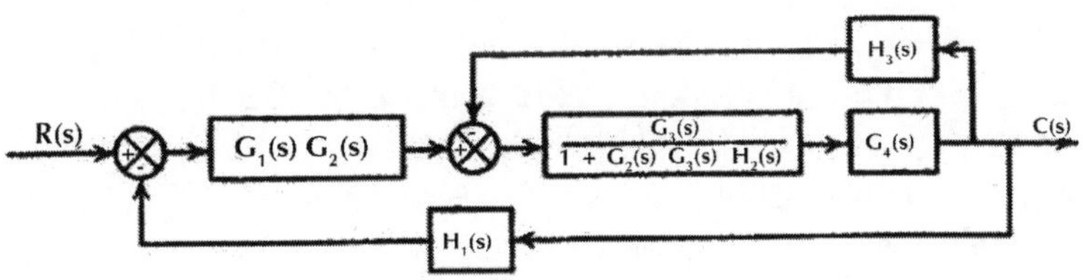

FIGURE 6 Reduced the equivalent intermediate loop.

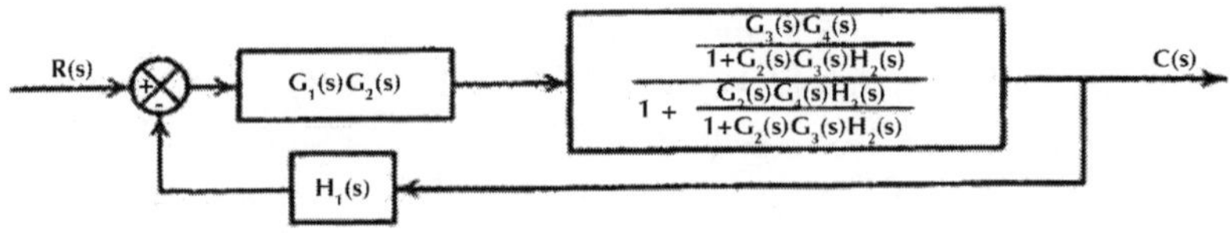

FIGURE 7 Reduced the equivalent minor loop.

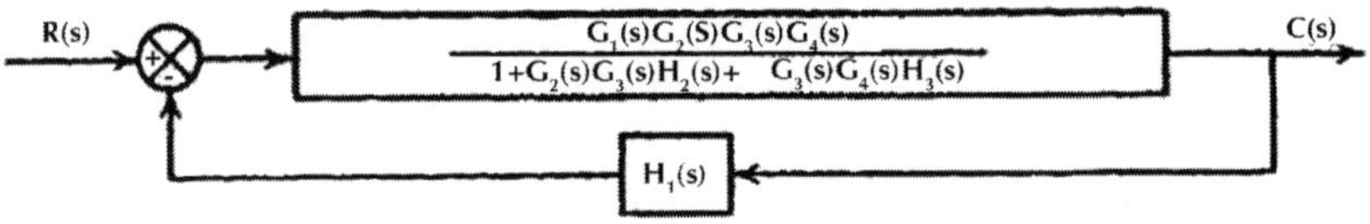

FIGURE 8 The equivalent feedback system.

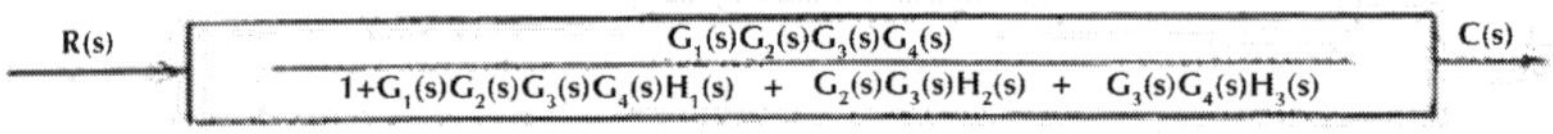

FIGURE 9 The system transfer function.

It becomes quite clear that such diagrams have many serious limitations of which the lack of ease, accuracy, and efficiency are the major flaws that can all but be eliminated by the use of signal flow diagrams.

A PHARMACOLOGICAL MODEL

In applying, a signal flow diagram to a pharmacological system of drug feedback it is important to have actual clinical data for source material.All of the data is from actual clinical and laboratory studies.

Each of the sections is divided into three test models. Model A that deals with delayed hyper sensitivity skin testing. Model B that is the Triple-Test Plan for Serology Diagnosis in Syphilis and Model C on the drug cyclosporine. Information of each of these sections that is drug descriptions, testing methods, and testing standards, are found in the appendix.

In Model A, the signal flow diagram graphs the rank by percentile of reactions to the six antigens of the hypersensitivity skin test.The sample size was 76 normal adult subjects and the following is the results of the six antigens, plus two agents that are also known indicators of possible anergy, Coccidioidin and Mumps, as represented in Table 1.

TABLE 1

Antigen	Intermediate Strength	Second Strength
Candidin	39%	92%
Coccidioidin	19%	45%
Mixed Respiratory Vaccine	41%	n/a
Mumps	78%	n/a
PPD	26%	83%
SK-SD	55%	93%
Staphage Lysate	71%	n/a
Trichophytin	28%	n/a

The rank of percentile can be indicated by the following signal flow diagram as represented in Figure 10.

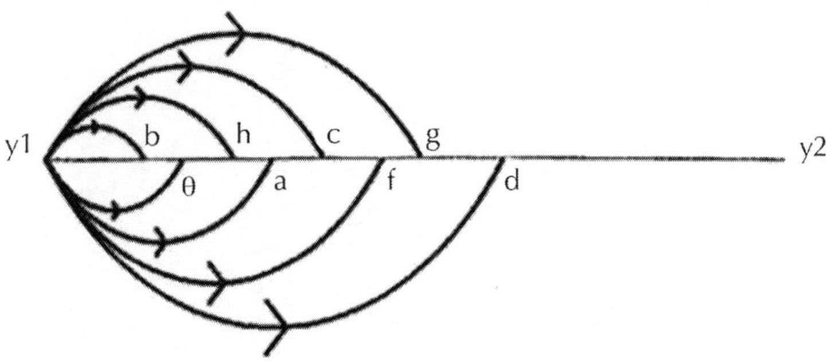

FIGURE 10 Signal flow diagram.

The following table, Table 2, represents the variables that is antigens, as they rank in percentile.

TABLE 2.

y_1 is the source.
a is Candidin at 39%
b is Coccidioidin at 19%
c is Mixed respiratory vaccine at 41%
d is Mumps at 78%
e is PPD at 26%
f is SK-SD at 55%
g is Staphage Lysate at 71%
h is Trichophytin at 28%
y_2 is the sink.

From this Table 2 and Figure 10, the best antigen for evaluating energy is Staphage Lysate, at 71%, with SK-SD following in second with 55%. The 78% rating for mumps is greater than Staphage Lysate but was not a part of the six antigens scheduled for use.

In Model B, the Triple-TestPlan for Serologic Diagnosis of Syphilis is taken from a flow chart in clinical laboratory diagnosis (Levinson and MacFate, 1969) and is represented by Figure 11.

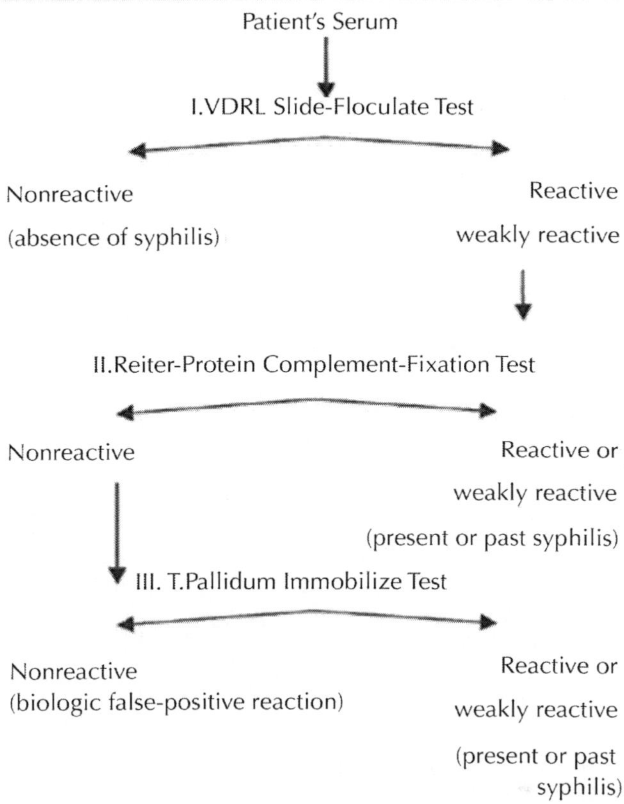

FIGURE 11 Flow chart ofTriple-Test Plan for Serologic Diagnosis of Syphilis.

This flow chart can be represented in a signal flow diagram as represented in Figure 12.

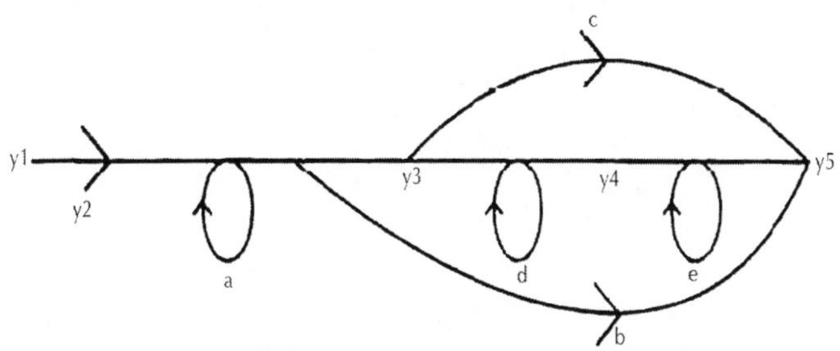

FIGURE 12 Signalflow diagram Triple-Test Plan for Serologic Diagnosis of Syphilis

The following is a table, Table 3, representing the itemsin Figure 12.

TABLE 3 Triple-test Plan for Serologic Diagnosis of Syphilis

y_1 is patient's serum and is the signal source.
y_2 is the VDRL Test.
a is the nonreactive response and is a feedback loop.
b is reactive and is a path.
y_3 is the Reiter-Protein Complement Test.
c is reactive and is a path.
d is nonreactive and is a feedback loop.
y_4 is T. Palladium Test.
e is nonreactive and is a feedback loop.
y_5 is reactive and is the sink.
The path gain is the line between y_1 and y_5 and y_2, y_3, and y_4 are the test variables.

Model C is the possible effects on laboratory tests using cyclosporine and is represented by the signal flow diagram in Figure 13.

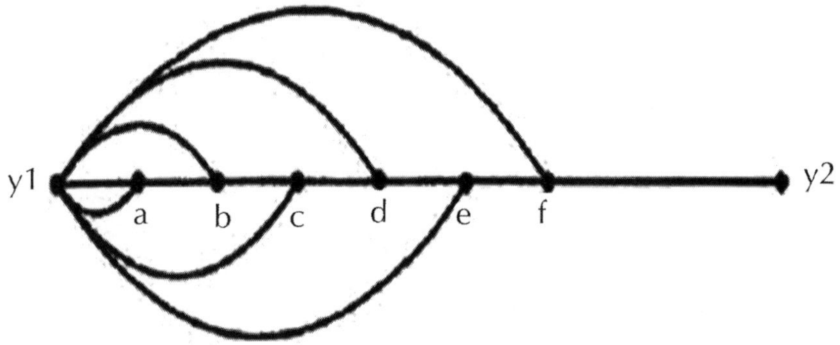

FIGURE 13 Signal flow diagramusing cyclosporine.

The following, table, Table 4, represents the items in Figure 13.

TABLE 4

y_1 is the source.
a is the blood cell count (decrease).
b is the blood potassium level (increase).
c is the blood uric acid level(increase).
d is the blood platelets, white cells, and magnesium (decrease).
e is the liver function tests (increase).
f is the kidney function tests (increase).
y_2 is the sink.
All of these conditions represent negative reactions to the laboratory tests[5].

DISCUSSION AND ANALYSIS OF RESULTS

In reviewing, the data from the three models of signal flow diagrams in the Pharmacological model in Chapter 6 of this dissertation, the difference between block diagrams and signal flow diagrams becomes strikingly apparent. In Model A, the signal flow diagram graphs the rank by percentile of reactions to the six antigens of the hypersensitivity skin test. In this model, the contrast is not so much between a table or chart of information but rather how such information would look in a signal flow diagram.

Upon first inspection of the two systems of representing information, it is apparent that the signal flow diagram is providing a specific type of information more quickly and more clearly than the table of information, mainly that the rank of percentile is more obvious by the use of branches and nodes on the signal flow diagram than is numerically or alphabetically represented in the table. This is a classic case of semanticallyverses visual information as the table is a linguistic and numerical device and the signal flow diagram is a visual or graphic device.

In taking into account the space needed to generate the amount of information, the signal flow diagram is superior in that it takes less space, in this case about one quarter of the space of the table, and has a clear directional sense, pointing arrows, and a correspondinghierarchy of branches andnodesrepresenting theindividual itemsandthere ranking by percent. In Model B, the Triple-Test Plan for Serologic Diagnosis of Syphilisis first represented by a flowchart and then by a signal flow diagram. The flow chart is a popular method of representing a process hierarchy and is found in most information oriented disciplines.

The flow chart is a visually clear representation of the information and affords more information than a table or a chart but as compared to the signal flow diagram, it is again overly complex and time consuming when matched with the simplicity of the signal flow diagram. Notice that the nodes denoting the non reactive response are clearly represented by loops and that the branches denoting a reactive response are all branched into there active sink node of the diagram.This is a more clear representation of the information than can be inferred from the flow chart and takes up less space than the flow chart even with the corresponding table of data.

Model C is the possible effects on laboratory tests using cyclosporine and is a diagram from data collected from a clinical case study of the drug.The signal flow diagram is more effective in describingand detailing the drug effects than the raw data and has the added benefit of clearly representing increases and decreases by the direction of the branch arrows. It is clear from these examples that signal flow diagrams are superior to tables, charts, flow charts, numerical, and word representations but it is also clear which signal flow diagrams are superior to other graphical representations most notably the formal or block diagram.

In the chapter signal flow verses block diagrams in this dissertation, the complexity, time, and space involved to represent equations in block diagrams becomes apparent as the more complex a system, the more involved becomes the block diagram until, it becomes a daunting task not only to design the block diagram but to read and comprehend the information represented with the diagrams. At a certain saturation point the block diagram becomes redundant as the information is no longer being imparted by this method of graphic representation. The block diagram reaches this saturation point long before the signal flow diagram does andis atest of the simplicity of the design of such a graphic representation.

From the point of view of the information sciences the signal flow diagram is a superior method of visually displaying mathematical equations into simple but accurate representations of information. Cybernetics is involved on one level in that the integration of information from raw data to usable information with complexity, time, and space being constraints to how the information is processed, clearly demonstrates the importance of simple communication systems such as the signal flow diagrams in increasing the efficiency and feedback response, on all levels, of the information sciences.

This is clearly a communication science as it is an information science in that mathematical, numerical, and word models are represented by this form of visual graphicdisplay and that the signal flow diagram is the superior form of this representation.

CONCLUSION

The use of signal flow diagrams in graphing feedback data for pharmacology has been shown to be of great importance in making the process of drug diagnosis more efficient, effective, and more accurate than either block diagrams or equations in processing complex systems. This method can substantially reduce staff burnout, costs, and need less errors that will have substantial positive effects on patient costs, wellbeing, and success rates.

The following can be said of the strengths of the signal flow diagram:

1. The signal flow diagram is a simplified graphic representation of mathematical, numerical, and word models and that these models are best expressed by the signal flow diagram.
2. The nodes and branches of the signal flow diagram can be a symbol of all types of data and information and the branches' arrows can represent loops, increases and decreases in relation to the information being represented and the nodes denoting a hierarchy of the information being represented.
3. Time and space are saved by the use of signal flow diagrams and this can be important when labor, cost, and efficiency factors are involved.
4. The complexity ofthe information is made simpler and more visually clear by the use of signal flow diagrams and that this simplicity is inherent in such a graphical method.
5. The signal flow diagram is superior to formal or block diagrams as block diagrams are inherently weak in the area of simplification and ease of use and are also time and space sensitive.
6. Both equations and raw data are inferior to signal flow diagrams in that equations are long, complex and time consuming and raw data is marginal at imparting specific information when compared to the signal flow diagram.
7. The signal flow diagram is superior to tables, charts, and flow charts in that it more readily accepts large quantities of information and represents the min the most accurate and simplistic manner, something that tables, charts, and flow charts perform with limited success.
8. The information saturation point of signal flow diagrams is higher than other forms of information representation.
9. Overall simplicity of conception, use, and understanding is the main point of interest and support for the signal flow diagram.
10. Accuracy ofthesignal flow diagram is in the simplicity of its use.The signal flow diagram is the graphical method of choice for the representation of mathematical, numerical, and word models and is superior to equations, raw data, tables, charts, flow charts, and block diagrams in representing the desired information.

RECOMMENDATIONS FOR FUTURE WORK

It is hoped that this study will inspire a more comprehensive program in the future and is a strong indicator of the successful application anduse of signal flow diagrams in pharmacology.

An interesting development of the signal flow diagram would be the color coding of the branches and nodes of the signal flow diagram to give greater representation and contrast to each of the branches and nodes. Another development would be the use of computer graphics to represent such diagrams and ease the making of such diagrams by having them integrated into the software programs.

A greater variety of data and information should be used in the future to give a wider representation of the uses of signal flow diagrams and the testing of other graphic and display oriented visual information. Such future research can only strengthen the use of such visual information for a wider use in all the information related disciplines.

APPENDIX A

A NEW FOUNDATION FOR INFORMATION

INTRODUCTION

The implementation of a ternary or quaternary based system to information infrastructure to replace the archaic binary system. Using a ternary or a quaternary based system will add greater robustness, compression, and utilizability to future information systems. With the advent of the superior compression of both a ternary and quaternary based system over that of the traditional binary system in information theory, the real need for a practical application to the fundamental structure of "information" must be re-considered for the 21st Century (Tice, 2006a and 2006b). With information technology being the"major driver" of economic growth in the past decade, adding $2 trillion ayear to the economy, the need to sustain and increase economic growth becomes an imperitive (Davies, 2007).

With a growing interest in"rebuilding"the internet, the fundamental question arises "why be tied to an archaic binary based system when both a ternary and a quaternary based system are more robust, offer greater utilizability, and have far greater capacity for compression? (Jesdanun, 2007 and Tice, 2006a and 2006b).Theanswer tothis question lies with the political aspect of the innovation process. If such a system is to be built using a ternary or a quaternary based system overthe out-dated binary based system then the government must be informed of the value of such systems over the existing system of information based infrastructures (Boehlert, 2006).

PART I

Information based systems use a binary based system represented by either a 1 or a 0. First developed by Claude Shannon in 1948 and termed "information theory", this fundamental unit has become the "backbone" of information age (Gates, 1995). One important aspect to information theory is that data compression, the removal of redundant features in a message, can "reduce" the overall size of a message (Gates, 1995). The need for better compression of messagesis an ever growing necessity in both computing and communications (Gates, 1995). The 2007 Nobel Prize for physics was awarded for the discovery of GMR that has increased the capacity of computer hard drives (Cho, 2007). A more profound effect to the computer industry would be the change from a binary based system to a ternary or quaternary based system.

The Internet was a by product of the"coldwar". A government sponsored project to develop a communications system that was decentralized (Nuechterlein and Weiser, 2005). Under the Department of Defense's Advanced Research Project Agency (DARPA), the Internet started life as ARPAnet in 1969 (Nuechterlein and Weiser, 2005). Even Tim Berners-Lee, the "father of the web", states that "The web is far from "done" and that it is a "jumbled state of construction" (Berners-Lee, 1999).

The outgrowth of a Ph.D. dissertation, Google, the search engine company, with perhaps the most extensive computing platformin existence, wants to become an information giant (Battelle, 2005). Google is in some respect a "Money Machine" with a value of $23 billion when it first hit the stock market in 2004 and has recorded an annual profit of $3 billion in 2006 (Vise and Malseed, 2005and Durman, 2007).

PART II

The advantages of a ternary and quaternary based system over a binary based system for information theory.

GROUP A

Binary non-random sequence
[111000111000111]

GROUP B

Binary random sequence
[111001100011111]

If group A and group B are compressed using the first character type and the following similar character types in a sequential order that follows the first character type, a numerical value to the number of character types can be assigned from the similar sequence of characters that can be represented by a multiple of number which represented in the group. An example will be that [111] equals the character type 1 multiplied by three to equal [111]. Notice that the character type is not a numerical one and does not have a semantic value beyond being different than the other character type [0].

Using a key code as an index of which character is to be multiplied, and by what amount, a compressed version of the original length of characters results.

KEY CODE A (GROUP A)

1 = x3
0 =x3

GROUP A

Binary non-random sequence
[10101]
Resulting in group A having a compression one third the original character length of 15 characters.

KEY CODE B (GROUP B)

1 = x3
0 = x3
Group B
Binary random sequence
[1_00110_11111]
Resulting in group B having a compression two thirds the original character length of 15 characters.

If a ternary system, or radix 3 based system, was used to represent both random and non-random sequential strings, the following three character symbols can be used:[1], [0], and [Q].

GROUP A

A non-random ternary sequence
[111000QQQ111000QQQ]
Total character length of 18 characters.

GROUP B

A random ternary sequence
[111000QQQQ11000QQQQ]
Total character length of 18 characters.

Again use of a key code to compress the original total character length by use of multiplication.

KEY CODE A (GROUP A)

1 = x3
0 = x3
Q = x3

GROUP A NON-RANDOM TERNARY SEQUENCE

[10Q10Q]
Total compression for group A is a length of six chracters from the original 18 character length. This is one third the original character length.

KEY CODE B (GROUP B)

1 = x3
0 = x3
Q = x4

GROUP B RANDOM TERNARY SEQUENCE

[1_0_Q_110_Q_]
Total compression for Group B is a length of seven characters from the original18 character length. This is less than one half of the original character length.

If a quaternary, or radix 4 based system, was used to represent both random and non-random sequential strings, the following character symbols can be used [1], [0], [Q], and [I].

GROUP A

A non-random quaternary sequence
[111000QQQIII111000QQQIII]
Total character length of 24 characters.

GROUP C

A randon quaternary sequence
[1110000QQQIII111100QQIII]

Total character length of 24 characters.

The use of a key code to compress the original character length by use of multiplication.

KEY CODE A (GROUP A)

1 =x3
0 = x3
Q= x3
I = x3
Group A non-random quaternary sequence.
[10QI10QI]

Total compression for group A is a length of eight characters from the original 24 character length. This one third the original character length.

KEY CODE C (GROUP C)

1 = x
0 = x
Q = x
I = x

GROUP C RANDOM QUATERNARY SEQUENCE

[1110_Q_I_1_OOQQI]

Total compression for group C is a lengthof 12 characters from the original 24 character length. This is one half the original character length.

PART III

In 2000, the "Milenium Bug", or Y2K problem, arose from the perceived problem of information systems changing from one century mark to another. The concern over this problem was global in scope. Imagin the entire information system of the world being made"redundant"bya superior information system? The concern I have for the United States is that a foreign power will implement a ternary or quaternary based information system that will "outdate" existing binary based systems.The reason for this book is to educate policy makers to the potential power of both a ternary and quaternary based information systems (Boehlert,2006).

APPENDIX B

COMPRESSION AND GEOMETRIC DATA

ABSTRACT

Kolmogorov complexity defines a random binary sequential string as being less patterned than a non-random binary sequential string. Accordingly, the non-random binary sequential string will retain the information about its original length when compressed, where as the random binary sequential string will not retain such information. In introducing a radix 2 based system to a sequential string of both random and non-random series of strings using a radix 2, or binary, based system.When a program is introduced to both random and non-random radix 2 based sequential strings that notes each similar subgroup of the sequential string as being a multiple of specific character and affords a memory to unit of information during compression, a sub-maximal measure of Kolmogorov complexity results in the random radix 2 based sequential string. This differs from conventional knowledge of the random binary sequential string compression values.

Traditional literature regarding compression values of a random binary sequential string have an equal measure to length that is not reducible from the original state (Kotz and Johnson, 1982). Kolmogorov complexity states that a random sequential string is less patterned than a non-random sequential string and that information about the original length of the non-random string will be retained after compression (Bradley). It is the result of the development of algorithmic information theory that was discovered in the mid-1960's (Solomonoff, 1964; Kolmogorov, 1965; Chaitin, 1969). Algorithmic information theory is a sub-group of information theory that was developed by Shannon in 1948 (Shannon, 1948).

Recent work by the author has introduced a radix 2 based system, or a binary system, to both random and non-random sequential strings (Tice, 2009). A patterned system of segments in a binary sequential string as represented by a series of 1's and 0's is rather a question of perception of subgroups within the string, rather than an innate quality of the string itself. While algorithmic information theory has given a definition of patterned verses patternless in sequential strings as a measure of random verses non-random traits, the existing standard for this measure for Kolmogorov complexity has some limits that can be redefined to form a new sub-maximal measure of Kolomogorov complexity in sequential binary strings (Kotz and Johnson, 1982). Traditional literature has a non-random binary sequential string as beingsuch: [111000111000111] resulting in total character length of 15 with groups of 1's and 0's that are sub-grouped in units of threes. A random binary sequence of strings will look similar to this example–[110100111000010] resulting in a mixture of sub-groups that seem "less patterned" than the non-random sample previously given.

The compression is the quality of a string to reduce from its' original length to a compressed value that still has the property of "decompressing" to its' original size without lose of theinformation inherent in the original state before compression.

This original information is the quantity of the strings original length before compression, bit length, as measured by the exact duplication of the l's and 0's found in original sequential string. The measure of the string's randomness is just a measure of the patterned quality found in the string.

The quality of "memory" of the original pre-compressed state of the binary sequential string has to do with the quantity of the number of l's and 0's in the string and the exact order of those digits in the original string are the measure of the ability to compress in the first place. Traditional literature has a nonrandom binary sequential string as being able to compress, while a random binary sequential string will not be able to compress. But if the measure of the number and order of digits in a binary sequence of strings is the sole factor for defining a random or non-random trait to a binary sequential string, then it is possible to "reduce" a random binary sequential string by some measure of itself in the form of sub-groups.

These sub-groups, while not being as uniform as a non-random subgroup of a binary sequential string, will nonetheless compress from the original state to one that has reduced the redundancy in the string by implementing a compression in each subgroup of the random binary sequential string. In other words, each sub-groupof the random binary sequential string will compress, retain the memory of that pre-compression state, and then, when decompressed, produce the original number and order to random binary sequential string.

The memory aspect to the random binary sequential string is, in effect, the retaining of the number and order of the information found in the original pre-compression state. This can be done by assigning a relation to the subgroup.It has a quality of reducing and then returning to the original state that can be done with the use of simple arithmetic. By assigning each subgroup in the random binary sequential string with a value of the multiplication of the amount found in the sub-group, a quantity is given that can be retained for use in reducing and expandingto the original size of quantity. It can be represented by a single character that represents the total number of characters found in the sub-group.

This is the very nature of compression and duplicates the process found in the non-random binary sequential strings. As an example the random binary sequential string [110001001101111] can be grouped into sub-groups as follows: {11}, {000}, {1}, {00}, {11}, {0}, and {111} with each sub-group bracketed into commonfamilies of like digits. An expedient method to reduce this string would be to take similar types and reduce to a single character that represented a multiple of the exact number of characters found in that sub-group. In this case taking the bracketed [11] and assign a multiple of 2 to a single character, one and then reduced it to a single character in the bracket that is underlined to note the placement of the compression. The compressed random binary sequential string would appear like this: [1000100101111] with the total character length of 13, exhibiting the loss of two characters due to the compression of the two similar sub-groups.

Decompression would be the removal of the underlining of each character and the replacement of the first characters to each of the sub-groups that would constitute a 100% retention of the original character number and order to the random binary sequential string. This makes for a new measure of Kolmogorov complexity in a random binary sequential string.

SUMMARY

The use of a viable compression method for sequential binary strings has applied aspects to transmission and storage of geometric data. Future book will explore practical applications to industry regarding applied aspects of compression to geometric data.

APPENDIX C

THE ANALYSIS OF BINARY, TERNARY, AND QUATERNARY BASED SYSTEMS FOR COMMUNICATIONS THEORY

ABSTRACT

The implementation of a ternary or quaternary based system to information infrastructure to replace the archaic binary system. Using a ternary or a quaternary based system will add greater robustness, compression, and utilizability to future information systems.

INTRODUCTION

With the advent of the superior compression of both a ternary and quaternary based system over that of the traditional binary system in information theory, the real need for a practical application to the fundamental structure of "information" must be reconsidered for the 21st Century. With information technology being the "major driver" of economic growth in the past decade, adding $2 trillion a year to the economy, the need to sustain and increase economic growth becomes an imperitive (Shannon, 1948).

With a growing interest in "rebuilding" the internet, the fundamental question arises "why be tied to an archaic binary based system when both a ternary and a quaternary based system are more robust, offer greater utilizability, and have far greater capacity for compression? (Tice, 2006; Li and Vitanyi, 1993/1997; Martin-Lot, 1996). The answer to this question lies with the political aspect of the innovation process. If such a system is to be built using a ternary or a quaternary based system over the out-dated binary based system then the government must be informed of the value of such systems over the existing system of information based infrastructures (Bradley).

PART I

Information based systems use a binary based system represented by either a 1 or a 0. First developed by Claude Shannon in 1948 and termed "information theory", this fundamental unit has become the "backbone" of information age. One important aspect to information theory is that data compression, the removal of redundant features in a message, can "reduce" the overall size of a message. The need for better compression of messages is an ever growing necessity in both computing and communications. The 2007 Nobel Prize for physics was awarded for the discovery of GMR that has increasedthe capacity of computer hard drives. A more profound effect to the computer industry would be the change from a binary based system to a ternary or quaternary based system.

The Internet was a by product of the "cold war".A government sponsored project to develop a communications system that was decentralized. Under the Department of Defense"s Advanced Research Project Agency (DARPA), the internet started life as ARPAnet in 1969. Even Tim Berners-Lee, the "father of the web", states that "the web is far from done"and that it is a "jumbled state of construction".

The outgrowth of a Ph.D. dissertation, Google, the search engine company, with perhaps the most extensive computing platform in existence, wants to become an information giant [13]. Google is in some respect a "Money Machine" with a value of $23 billion when it first hit the stock market in 2004 and has recorded an annual profit of $3 billion in 2006.

PART II

The advantages of a ternary and quaternary based system over a binary based system for information theory.

RADIX 2 BASE

Radix 2 Based System

GROUP A

Binary non-random sequence
[111000111000111]

GROUP B

Binary random sequence
[111001100011111]

If Group A and group B are compressed using the first character type and the following similar character types in a sequentialorder that follows the first character type, a numerical value to the number of character types can be assigned from the similar sequence of characters that can be represented by a multiple of number represented in group. An example will be that [111] equals the character type 1 multiplied by three to equal [111]. Notice that the character type is not a numerical one and does not have a semantic value beyond being different than the other character type [0].

Using a key code as an index of which character is to be multiplied, and by what amount, a compressed version of the original length of characters results.

KEY CODE A (GROUP A)

1 = x3
0 =x3

GROUP A

Binary non-random sequence
[10101]
Resulting in group A having a compression one third the original character length of 15 characters.

KEY CODE B (GROUP B)

1 = x3
0 = x3

GROUP B

Binary random sequence
[1_00110_11111]
Resulting in group B having a compression two thirds the original character length of 15 characters.

A RADIX 3 BASE

RADIX 3 BASED SYSTEM

If a ternary system, or radix 3 based system, was used to represent both random and non-random sequential strings, the following three character symbols can be used [1], [0], and [Q].

GROUP A

A Non-random ternary sequence
[111000QQQ111000QQQ]
Total character length of 18 characters.

GROUP C

A Random ternary sequence
[111000QQQQ11000QQQQ]
Total character length of 18 characters.

Again use of a key code to compressthe original total character lengthby use of multiplication.

KEY CODE A (GROUP A)

1 = x3
0 = x3
Q = x3
Group A non-random ternary sequence
[10Q10Q]

Total compression for Group A is a length of 6 chracters from the original 18 character length. This is one third the original character length.

KEY CODE C (GROUP C)

1 = x3
0 = x3
Q = x4
Group C random ternary sequence
[1_0_Q_110_Q_]

Total compression for group C is a length of sevencharacters from the original 18 character length. This is less than one half of the original character length.

RADIX 4 BASE

RADIX 4 BASED SYSTEM

If a quaternary, or radix 4 based system, was used to represent both random and non-random sequential strings, the following character symbols can be used [1], [0], [Q], and [I].

GROUP A

A Non-random quaternary sequence
[111000QQQIII111000QQQIII]
Total character length of 24 characters.

GROUP D

A Randon quaternary sequence.
[1110000QQQIII111100QQIII]
Total character length of 24 characters.

The use of a key code to compress the original character length by use of multiplication.

KEY CODE A (GROUP A)

1 = x3
0 = x3
Q= x3
I = x3
Group A non-random quaternarysequence.
[10QI10QI]

Total compression for group A is a length of eight characters from the original 24 character length. This one third the original character length.

KEY CODE D (GROUP D)

1 =x
0=x
Q=x
I =x
Group D randomquaternarysequence
[1110_Q_I_1_00QQI_]

Total compression for group Dis a length of 12 charactersfrom the original 24 character length.This is one half the original characterlength.

PART III

In 2000, the "Milenium Bug", or Y2K problem, arose from the perceived problem of information systems changing from one century mark to another. The concern over this problem was global in scope. lmagin the entire information system of the world being made "redundant" by a superior information system? The concern I have for the United States is that a foreign power will implement a ternary or quaternary based information system that will "outdate" existing binary based systems. The reason for this book is to educate policy makers to the potential power of both a ternary and quaternary based information systems.

SUMMARY

The results of using a compression engine to compress both random and non-random sequential strings of radix 2, radix 3, and radix 4 based strings resulted in the following:

Radix Base		Random	Non-random
Radix 2	15 character length total	11	5
Radix 3	18 character length total	7	6
Radix 4	24 character length total	12	8

Both the radix 3 and radix 4 based systems had substantial compression values in the random sequential strings categories. As random sequential strings have the most applicable nature to practical modes of information transmission and storage, these findings have both theoretical and applied aspects to communication theory in all of its manifestations.

APPENDIX D

THE USE OF A RADIX 5 BASE FOR TRANSMISSION AND STORAGE OF INFORMATION

ABSTRACT

The radix 5 based system employs five separate characters that have no semantic meaning except not representing the other characters. Traditional literature has a random stringofbinary sequential charactersas being "less patterned" than non-random-sequential strings. A non-randomstring of characters will be able to compress, were as a random string of characters will not be able to compress. This study has found that a radix 5 based character length allows for equal compression of random and non-random sequential strings. This has important aspects to information transmission and storage.

INTRODUCTION

As communications handle an ever growing amount of information for transmission and storage, the very real need for an upgrade in the fundamental structure of such a system has come to light. As the very bases of coding is compression, the greater the amount of information compressed, the more efficient the system. The earliest calculating machine as the human hand, it is five digits representinga natural symmetry found, with frequency, in the organic world (Ifrah, 2000 and Weyl). A radix 5 base system, also known as a quinary numeral system, is composed of five separate characters that have no meaning apart from the fact the each character is different than the other characters.This is a development from the binary system used in Shannon"s information theory (Shannon, 1948).

PART I

The radix 5 base is not the traditionalbinary based error-detection and error-correcting codes that are also known as "prefix codes" and use a 5-bit length for decimal coding

(Richards, 1995). A radix 5 base is composed of five separate symbols with each an individual character with no semantic meaning. A random string of symbols has the quality of being "less patterned" than a non-random string of symbols. Traditional literature on the subject of compression, the ability for a string to reduce in size while retaining"information" about its original charactersize, states that a non-randomstring of characters will be able to compress, were as the random string of characterswill not compress (Kotz and Johnson, 1982).

PART II

The following examples will use the following symbols for a radix 5 based system of characters (Example A)

EXAMPLE A

0
O
Q
1
I

The followingis an example of compression of a random and non-random radix 5 base system. A non-random string of radix 5 based characters with a total 15 character length [Group A].

Group A 000OOOQQQ111III

A random string of a radix 5 based characters with a total of 15 character length [Group B].

GroupB 000OOQQQQ11IIIII

If a compression program were to be used on group A and group B that consisted of underlining the first individual character of a similar group of sequential characters, moving towards the right, on the string and multiplying it by a formalized system of arithmeticas found in a key, see key code A and key code B, with the compression of group A and group B as the final result.

KEY CODE A (FOR GROUP A)

0 = X3
O = x3
Q = x3
1 = x3
I = x3

Group A 0O Q 1 I
Resulting in a five character length for group A.

KEY CODE (FOR GROUP B)

0 =x3
Q =x2
Q =x4
L =x2
I = X4
Group B oOQ1I
Resulting in a five character length for Group B.

Both group A (non-random) and group B (random) have the same compression values, each group resulted in a compression value of 1/3 the total pre-compression, original, state. This contrasts traditional notions of random and non-random strings (Kotz and Johnson, 1982). These findings are similar to Tice (2003) and have applications to both algorithmic information theory and information theory (Tice, 2003).

Some other examples using example A radix five characters [0OQ11] to test random and non-randomsequential strings.

The following is a non-random string of a radix five based characters with a total of 15 character length (GroupA).

Group A 000OOOQQQ111III

A random string of a radix five based characters with a total of 15 character length (Group C).

Group C 00000OQQQQQ1I

If a compression program were to be used on group A and group C that consist of underlining the first individual character of a similar group of sequential characters, moving towards the right, on the string and multiplyingit by a formalized system of arithmetic as found in a key, see key code A and key code C, with the compression of group A and group C as the final result.

KEY CODE C (FOR GROUP A)

0 = x3
O = x3
Q = x3
1= x3
1 = x3
Group A 0OQ1I
Resulting in a five character length for Group A.

KEY CODE C (FOR GROUP C)

O = x5
O = xl
Q = x5
1 = x1
I = x1
Groupe C oOQlI

Resulting in a five character length for Group C.

This example has group A as a non-random string and group D as a random string using radix five characters for a total 15 character length.

A non-random string of radix five characters with a 15 character length (Group A).

Group A 000OOOQQQ111III

A random string of a radix five based characters with a total of 15 character length (Group D).

Group D 0OOOOQQ1111IIII

If a compression program were to be used on group A and group D that consisted of underlining the first individual character of a similar group of sequential characters, moving towards the right, on the string and multiplyingit by a formalized system of arithmetic as found in a key, see key code A and key code D, with the compression of group A and group D as the final result.

KEY CODE A (FOR GROUP A)

0 = x3

O = x3

Q = x3 1 = x3

I = x3

Group A 0OQ1I

Resulting in a five character length for Group A.

KEY CODE D(FOR GROUP D)

0 = xl

O = x4

Q = xl

1 = x4

I = x4

Group D 0OQ1I

Resulting in a five character length for group D.

As a final example group A is a non-random sequential string and group E as a random sequential string using a radix fivecharacters for a total of 15 character length.

A non-random string of radix five based characters with a 15 character length (Group A).

Group A 000OOOQQQlllIII

A random string of a radix five based characters with a total of 15 character length (Group E).

GroupE 00OOOOQQQQlllII

If a compression program were to be used on group A and group E that consisted of underlining the first individual character of a similar group of sequential characters, moving to the right, on the string and multiplying it by a formalized system of arithmeticas found in a key, see key code A and key code E, with the compression of group A and group E as the final result.

KEY CODE A (FOR GROUP A)

0 = x3
O = x3
Q = x3
l = x3
I = x3
Group A 0OQlI
Resulting in a five character length for group A.

KEY CODE E (FOR GROUP E)

0 = x2
O = x4
Q = x4
L = x3
I = x2
Group E oOQlI
Resulting in a five character length for group E.

Again, these examples conflict with traditional notions of random and non-random sequential strings in that the compression ratio is one third of the original character number length for both the random and non-random sequential strings using a radix 5 base system.

PART III

Traditional information based systems use a binary based system represented by either a1 or a 0. First developed by Claude Shannon in 1948 and termed "information theory", this fundamental unit has become the backbone of information age. One important aspect to information theory is that data compression, the removal of redundant features in a message that canreduce the overallsize of a message (Gates, Myhrvold, and Rinearson). With the substantial compression values found in using a radix 5 based system. It seems a new paradigm has arrived to carry the future of information.

Information technology has been the major driver of the economic growth in the past decade adding $2 trillion a year to the economy (Davis, 2007). This growth needs to be sustained in order for new jobs and the economy to maintain a high standard of living. Only by considering alternative developments to existing models of technology, can the future of the economy develop and continue at a successful level of growth.

The Internet was an outgrowth of the cold war as a governmentsponsored project to develop a communications network that was decentralized (Nuecherlein and Weiser, 2005). Today, the internet is the major highway of global information with search engine technology rapidly taking center stage on both universities research departments as well as the Dow Jones index. The need to handle this vast and ever growing amount of information will need a fundamental change to the very nature of the structure of information systems. It is clear that any new developments to deal with more and more information must begin at the fundamental level.

With a radix 5 base that has been proved to have the compression ratio similar in both random and non-random states, the question of usage as a medium for transmission and storage of information becomes para mount. With an ever increasing need for transmission and storage in the areas of telecommunications and computer science, the viability of a new system at the fundamental level of communication theory that is both robust and diverse enough to allow for future growth beyond the binary based system in use today.

SUMMARY

This appendix has shown that a radix 5 based system has profound properties of compression that are well beyond those found in binary systems using sequential strings of a random and non-random types. These compression values have strong potential applications to information theory and communication theory as a whole.

While the identical compression values for random and non-random radix 5 based strings is a result of this appendix, the application of this theory to communication theory cannot be understated. It has been shown that a radix 5 based system has a compression factor that makes it an ideal functional standard for future information systems, particularly in the fields of telecommunications and computer science.

APPENDIX E

A COMPARISON OF A RADIX 2 AND A RADIX 5 BASED SYSTEMS

ABSTRACT

A radix 2 based system is composed of two separate character types that have no meaning except not representing the other character typeas defined by Shannonin 1948. The radix 5 based system employs five separate characters that have no semantic meaning except not representing the other characters. Traditional literature has a random string of binary sequential characters as being "less patterned" than non-random sequential strings. A non-random string of characters will be able to compress, were as a random string of characters will not be able to compress. This study has found that a radix five based character length allows for equal compression of random and non-randomsequential strings. This has important aspects to information transmission and storage.

INTRODUCTION

As communications handle an ever-growing amount of information for transmission and storage, the very real need for an upgradein the fundamental structure of such a system has come to light. As the very bases of coding is compression, the greater the amount of information compressed, the more efficient the system. The earliest calculating machinewas the humanh and, its five digits representing a natural symmetry found, with frequency, in the organic world (Ifrah, 2000 and Weyl, 1956). A radix 5 base system, also known as a quinary numeral system, is composed of five separate characters that have no meaning apart from the fact the each characteris different than the other characters. This is a development from the binary system used in Shannon's information theory (Shannon, 1948).

THE RADIX 2 BASED SYSTEM

The radix 2 based system is a two character system that has no semantic meaning except not representing the other character type. The traditional 1 and 0 will be used in this Appendix.

The following is an example of compression of a random and non-random radix 2 based systems. A non-random sequential stringof characters will have a total length of 15 characters as seen in group A.

Group A: 111000111000111

A random sequential string of characters will have a total length of 15 characters as seen in groupB.

Group B: 110000111110111

If a compression program were to be used on group A and group B that consisted of underlining the first individual character of a similar group of sequential characters, moving towards the right, on the string and multiplying it by a formalized system of arithmeticas found in a key, see key code 1 and key code 2, with the compression of group A and group B as a formal result.

KEY CODE 1 (FORGROUP A)

L=x3

0=x3

Group A 10101

Resulting in a compressed state of five characters for Group A.

KEY CODE 2 (FOR GROUP B)

l=x5

0=x4

Group B 11010111

Resulting in a compressed state of eight characters for group B.

Compression values of the non-random binary sequential string are one third the original 15 character length and the random binary sequential string are almost half of the original random 15 character length.

PART I

The radix 5 base is not the traditional binary based error-detection and error correcting codes that are also known as "prefix codes" that use a 5-bitlength for decimal coding (Richards, 1955). A radix 5 base is composed of five separate symbols with each an individual character with no semantic meaning. A random string of symbols has the quality of being "less patterned" than a non-randomstring of symbols. Traditional literature on the subject of compression, the ability for a string to reduce in size while retaining"information" about its original charactersize, states that a non-random string of characters will be able to compress, were as the random string of characters will not compress (Kotz and Johnson 1982).

PART II

The following examples will use the following symbols for a radix 5 based system of characters (Example A).

Example A

0
O
Q
1
I

The following is an example of compression of a random and non-random radix 5 base system. A non-randomstring of radix 5 based characters with a total 5 character length (Group A).

Group A 000OOOQQQ111III

A random string of a radix 5 based characters with a total of 15 character length (Group B). Group B 000OOQQQQlIIIII

If a compression programwere to be used on group A and group B that consisted of underlining the first individual character of a similar group of sequential characters, moving towards the right, on the string and multiplying it by a formalized system of arithmeticas found in a key, see key code A and key code B, with the compression of group A and group B as the final result.

KEY CODE A (FOR GROUP A)

0 = x3
O = x3
Q = x3
1 = x3
I = x3
Group A 0OQ1I
Resulting in a five character length for group A.

KEY CODE B (FOR GROUP B)

0 = x3
O = x2
Q = x4
1 = x 2
I = X4
GroupB 0OQ1I
Resulting in a five character length for group B.

Both group A (Non-random) and group B (Random) have the same compression values, each group resulted in a compression value of 113 the total pre-compression, original, state. This contrasts traditional notions of random and non-random strings (Tice, 2003). These findings are similar to Tice (2003) and have applications to both algorithmic information theory and information theory (Gates, Myhrvold, and Rinearson, 1995)

Some other examples using example A radix five characters [0OQlI] to test random and non-random sequential strings.

The following is a non-random string of a radix 5 based characters with a total of 15 character length (Group A).

Group A 000OOOQQQ1lIII

A random string of a radix 5 based characters with a total of 15 character length (Group C).

Group C 00000OQQQQ1I

If a compression program were to be used on group A and group C that consist of underlining the first individual character of a similar group of sequential characters, moving towards the right, on the string and multiplying it by a formalized system of arithmetic as found in a key, see key code A and key code C, with the compression of group A and group C as the final result.

KEY CODE A (FOR GROUP A)

0 = x3

O = x3

Q = x3

1 = x3

l = x3

Group A 0OQ1I

Resulting in a five character length for group A.

KEY CODE C (FOR GROUP C)

0 =x5

O=x1

Q=x5

1=x1

I=x1

Group C oOQII

Resulting in a five character length for group C.

This example has group A as a non-random string and group D as a random string using radix five characters for a total15 character length.

A non-random string of radix 5 characters with a 15 character length (Group A).

Group A 000OOOQQQ111III

A random string of a radix 5 based characters with a total of 15 character length (Group D).

Group D oOOOOQQ1111IIII

If a compression program were to be used on group A and group D that consisted of underlining the first individual character of a similar group of sequential characters, moving towards the right, on the string and multiplying it by a formalized system of arithmetic as found in a key, see key code A and key code D, with the compression of group A and group D as the fmal result.

KEY CODE A (FOR GROUP A)

0 = x3

O = x3

Q = x3
1 = x3
I = x3
Group A oOQlI
Resulting in a 5 character length for Group A.

KEY CODE D (FOR GROUP D).

0 = x1
O = x4
Q = x1
1 = x4
I = x4
Group D oOQlI
Resulting in a 5 character length for group D.

As a final example group A is a non-random sequential string and group E as a random sequential string using a radix five characters for a total of 15 character length.

A non-random string of radix 5 based characters with a 15 character length (Group A).

Group A oooOOOQQQ111III

A random string of a radix 5 based characters with a total of 15 character length (Group E).

GroupE ooOOOOQQQ111II

If a compression program were to be used on group A and group E that consisted of underlining the first individual character of a similar group of sequential characters, moving to the right, on the string and multiplying it by a formalized system of arithmetic as found in a key, see key code A and key code E, with the compression of group A and group E as the formal result.

KEY CODE A (FOR GROUP A)

0 = x3
O = x3
Q = x3
1 = x3
I = x3
Group A 0OQlI
Resulting in a 5 character length for Group A.

KEY CODE E (FOR GROUP E)

0 = x2
O = x4
Q = x4
1 = x3
I = x2

Group E oOQlI

Resulting in a 5 character length for group E.

Again, these examples conflict with traditional notions of random and non-random sequential strings in that the compression ratio is one third of the original character number length for both the random and non-random sequential strings using a radix 5 base system.

PART III

Traditional information based systems use a binary based system represented by either a1 or a 0. First developed by Claude Shannon in 1948 and termed "information theory", this fundamental unit has become the backbone of our information age. One important aspect to information theory is that of data compression, the removal of redundant features in a message that can reduce the overall size of a message (Davis, 2007).With the substantial compression values found in using a radix 5 based system it seems a new paradigm has arrived to carry the future of information.

Information technology has been the major driver of the economic growth in the past decade adding $2 trillion a year to the economy (Nuecherleinand Weiser, 2005). This growth needs to be sustained in order for new jobs and the economy to maintain a high standard of living. Only by considering alternative developments to existing models of technology, can the future of the economy develop and continue at a successful level of growth.

The Internet was an outgrowth of the cold war as a government-sponsored project to develop a communications network that was decentralized. Today, the internet is the major highway of global information with search engine technology rapidly taking center stage on both universities research departments as well as the Dow Jones index. The need to handle this vast and ever growing amount of information will need a fundamental change to the very nature of the structure of information systems. It is clear that any new developments to deal with more and more information must begin at the fundamental level.

With a radix 5 base that has been proved to have the compression ratio similar in both random and non-random states, the question of usage as a medium for transmission and storage of information becomes paramount. With an ever increasing need for transmission and storage in the areas of telecommunications and computer science, the viability of a new system at the fundamental level of communication theory that is both robust and diverse enough to allow for future growth beyond the binary based system in use today.

SUMMARY

This appendixhas shown that a radix 5 based system has profound properties of compression that arewell beyond those found in binary systems using sequential strings of a random and non-random types. These compression values have strong potential applications to information theory and communication theory as a whole.

When comparing the radix 2 and the radix 5 based systems the greater compression factor of the radix 5 based system has strong applications to signal transmission and storage issues.

While the identical compression values for random and non-random radix 5 based strings is a result of this appendix, the application of this theory to communication theory cannot be understated. It has been shown that a radix 5 based system has a compression factor that makes it an ideal functional standard for future information systems, particularly in the fields of telecommunications and computer science.

APPENDIX F

RANDOM AND NON-RANDOM SEQUENTIAL STRINGS USING A RADIX 5 BASED SYSTEM

Kologorov complexity defines a random binary sequential string as being less patterned than a non-random binary sequential string. Accordingly, the non-random binary sequential string will retain the information about its original length when compressed, where as the random binary sequential string will not retain such information. In introducing a radix 5 based systems to a sequential string of both random and non-random series of strings using a radix 5, and quinary, based system. When a program is introduced to both random andnon-random radix 5 based sequential strings that notes each similar subgroup of the sequential string as being a multiple of that specific character and affords a memory to that unit of information during compression, a sub-maximal measure of Kologorov Complexity results in the random radix 5 based sequential string. This differs from conventional knowledge of the random binary sequential string compression values.

Traditional literatures regarding compression values of a random binary sequential string have an equal measure to length that is not reducible from the original state (Kotz and Johnson, 1982). Kologorovcomplexity states that a random sequential string is less patterned than a non-random sequential string and information about the original length of the non-random string will be retained after compression (Abide). Kologorov complexity is the result of the development of algorithmic information theory that was discovered in the mid-1960's (Solomonoff, 1969). Algorithmic information theory is a sub-group of information theory that was developed by Shannon in 1948 (Shannon, 1948).

Recent work by the author has introduced a radix 5 based system, or a quinary system, to both random and non-random sequential strings (Tice, 2008). A patterned system of segments in a binary sequential string as represented by a series of 1's and 0's is rather a question of perception of subgroups within the string, rather than an innate quality of the string itself. While algorithmic information theory has given a definition of patterned verses patternless in sequential strings as a measure of random verses non-random traits, the existing standard for this measure for Kolmogorov complexity has some l its' that can be redefined to form a new sub-maxim al measure of Kolomogorov complexity in sequential binary strings (Kotz and Johnson, 1982). Traditional literature has a non-random binary sequential string as beingsuch: [111000111000111] resulting in total character length of 15 with groups of 1's and 0's that are sub-grouped in units of threes. A random binary sequence of strings will look similar to this example: [110100111000010] resulting in a mixture of sub-groups that seem "less patterned" than the non-random sample previously given.

Compression is the quality of a string to reduce from its' original length to a compressed value that still has the property of "decompressing" to its' original size without the loss of theinformation inherent in the original state before compression. This original information is the quantity of the strings original length before compression, bit length, as measured by the exact duplication of the l's and 0's found in that original sequential string. The measure of the string's randomness is just ameasure of the patterned quality found in the string.

The quality of "memory" of the original pre-compressed state of the binary sequential string has to do with the quantity of the number of l's and 0's in that string and the exact order of those digits in the original string are the measure of the ability to compress in the first place. Traditional literature has a non-random binary sequential string as being able to compress, whilea random binary sequential string will not be able to compress. But if the measure of the number and order of digits in a binary sequence of strings is the sole factor for defining a random or non-random trait to a binary sequential string, then it is possible to "reduce" a random binary sequential string by some measure of itself in the form of sub-groups. These sub-groups, while not being as uniform as a non-random sub-group of a binary sequential string, will nonetheless compress from the original state to one that has reduced the redundancy in the string by implementing a compression in each sub-group of the random binary sequential string. In other words, each sub-group of the randombinary sequential string will compress, retain the memory of that pre-compression state, and then, when decompressed, produce the original number and order to random binary sequential string.

The memory aspect to the random binary sequential string is, in effect, the retaining of the number and order of the information found in the original pre-compression state. This can be done by assigning a relation to the subgroup that has a quality of reducing and then returning to the original state, that can bedone with the use of simple arithmetic. By assigning each sub-group in the random binary sequential string with a value of the multiplication of the amount found in that sub-group, a quantity is given that can be retained for use in reducing and expandingto the original size of the quantity and can be represented by asingle character that represents the total number of characters found in that sub-group. This is the very nature of compression and duplicates the process found in the non-random binary sequential strings. As an example the random binary sequential string [110001001101111] can be grouped into sub-groups as follows: {11}, {000}, {1}, {00}, {11}, {0}, and {111} with each sub-group bracketed into common families of like digits. An expedient method to reduce this string would be to take similar types and reduce to a single character that represented amultiple of the exact number of characters found in that sub-group. In this case taking the bracketed {11} and assign a multiple of 2 to a single character,1, and then reduced it to a single character in the bracket that is underlined to note the placement of the compression.

The compressed random binary sequential string would appear like this: [1000100101111] with the total character length of 13, exhibiting the loss of two characters due to the compression of the two similar sub-groups. Decompression would be the removal of the underlining of each character and the replacement of the1's characters to each of the sub-groups that would constitute a100% retention of the original character number and order to the random binary sequential string. This makes for a new measure of Kolmogorov complexity in a random binary sequential string.

This same method of compression can be used with a radix 5 based system that provides for an even greater measure of reduction than is found in the binary sequential string. The radix 5 base number system has five separate characters that have no semantic meaning except not representing the other characters in the five character system. The following five numbers will represent thefive characters found in the radix 5 base number system that will be used as an example in this appendix [0, 1, 2, 3 and 4]. As an example of a random radix 5 sequential string the following wouldappear like this: [001112233334440111223444] with a total character length of 24 characters. If all the applicable similar sequential three character's are compressed to a single representative character that represents the other two characters in the three character compressed unit of the string, then the following would result: [0012233334012234].

The underlined characters represent the initial position of the three character group of similar characters with a compressed state of 16 characters total. This is a reduction of one third the total original character length of 24 characters. A nonrandom radix 5 base sequential string will have the same character types: [0, 1, 2, 3,and4].But with a regular pattern of groupings such as [00112233440011223344] that has a total character length of 20 and if all two sequentially similar characters are compressed using all fivecharacter types thefollowing will occur: [0123401234] resulting in a compressed nonrandom radix 5 base sequential string of 10.

The appendixhas shown that a sub-maximal measure of Kolmogorov complexity exists that has implications to a new standard of the precise measure of randomness in both a radix 2 and a radix 5 based number systems.

APPENDIX G

A COMPARISON OF COMPRESSION VALUES OF BINARY AND TERNARY BASE SYSTEMS

ABSTRACT

The appendix will introduce the ternary, or radix 3, based system for use as a fundamental standard beyond the traditional binary, or radix 2, based system in use today. A compression level is noted that is greater than the known Martin-Lof standard of randomness in both binary and ternary sequential strings.

INTRODUCTION

A ternary, or radix 3, based system is defined as three separate characters, or symbols, that have no semantic meaning apart from not representing the other characters. This is the same notion Shannon gave to the binary based system used in his appendix on information theory upon it's publication in 1948 (Shannon, 1948). Richards has noted that the radix 3 based system as the most efficient base, more so than even the radix 2 or radix 4 based systems (Richards, 1955). A compression level is noted in this appendix that is greater than the known Martin-Lof standard of randomness in both binary and ternary sequential strings.

RANDOMNESS

The earliest definition for randomness in a string of 1's and 0's was defined by von Mises, but it was Martin-Lof"s paper of 1966 that gave a measure to randomness by the patterlessness of a sequence of 1's and 0"s in a string that could be used to define a random binary sequence in a string (Martin-Lof, 1966). This is the classical measure for Kolmogorov complexity, also known as algorithmic information theory, of the randomness of a sequence found in a binary string (Kotz and Johnson, 1982). Martin-

Lof (1966) also defined a random binary sequential string as being unable to compress from its original state. Non-random binary sequential strings can compress to less than there original state (Martin-Lof, 1966).

COMPRESSION PROGRAM

The compression program to be used has been termed the modified symboic space multiplier program as it simply notes the first character in a line of characters in a binary sequence of a string and sub-groups them into common or like groups of similar characters, all 1's grouped with 1's and all 0's grouped with 0's, in that string and is assigned a single character notation whichrepresents the number found in that sub-group, so,it can be reduced, compressed, and decompressed, expanded, back to it's original length and form. Anunderlined 1 or 0 is usually used to note the notation symbol for the placement and character type in previous applications of this program. An italicized character will be used for this.

BINARY SYSTEM

The binary system, also known as a radix 2 based system, is composed of two characters, usually a 0 and a 1, that have no semantic properties except not representing the other. Group A will represent a non-random sequential binary string and group B will represent a random sequential binary string. Both group A and group B will be 15 characters in total length.

Group A: [000111000111000] (Non-random)

Group B: [001110110011100] (Random)

Utilizing, the modified symbolic space multiplier program to processlike sequential characters, either 0's or 1's, into sub-groups and note them with an italicized character specific to that sub-group and having it represent a specific multiple of the sub-group as found in a key, in this case group A key and group B key, as a compressed aspect to both group A and group B sequential binary strings.

Group A Key: All italicized characters will represent a multiple of3.

Group B Key: The italicized character 0 will represent a multiple of 2 and the italicized character 1 will represent a multiple of 3.

Group A: [01010]

Group B: [01011010]

The compressed state of group A, non-random, is five characters in length. The compressed state of group B, random, is 8 characters in length. Note that the random sequential binary string in group B compressed to less than the original total pre-compression length. This differs from standards known in Martin-Lof randomness and those found in Kolmogorov complexity.

TERNARY SYSTEM

A ternary, or radix 3, based system there are three characters used that have no semantic meaning except not representing the other two characters. Group C will represent a non-random ternary sequential string and group D will represent a random ternary sequential string. The total length for each group, group C and group D will be 12 characters in length. The three characters to be used in this study are a 0, 1, and 2.

Group C: [001122001122]

Group D: [001222011222]

Again each group will be assigned a specific compression multiple based on a specific character type, in this case an italicized 0,1, and 2, as defined in a key, group C key and group D key.

Group C Key: The italicized characters 0, 1, and 2 will represent each a multiple of 2.

Group D Key: The italicized character 0 will represent a multiple of 2.The italicized character 1 will represent a multiple of 2 and the italicized character 2 will represent a multiple of 3.

Group C: [012012]

Group D: [012012]

The compressed state of group C, non-random, is 6 characters in length. The compressed string of Group D, random, is 6 characters in length. Again note that Group D, the random sequential ternary string, is less than it is pre-compressed state, and again, is novel for those extrapolations of binary examples found in Kolmogorov complexity.

APPLICATION OF THEORY

The compression of data for transmission and storage of information is the most practical application of a binary system in telecommunications and computing. The application of a ternary or radix 3 based systems to existing communication systems has many advantages. The first is the greater amount of compression from this base, as opposed to the standard binary based system in use today, of random strings of data, and secondly, as a more utilizable system because of the three characters, or symbol, based system that provides for more variety to develop information applications. From telecommunicationsto computing, the ternary based system applied at a fundamental standard would allow for a more robust communications system than is currently used today.

APPENDIX H

PATTERNS WITHIN PATTERNLESS SEQUENCES

While Kolmogorov complexity, also known as algorithmic information theory, defines a measure of randomness as being patternless in a sequence of a binary string, such rubrics come into question when sub-groupings are used as a measure of such patterns in a similar sequence of a binary string. This appendix examines such sub-group patterns and finds questions raised about existing measures for a random binary string.

Qualities of randomness and non-randomness have their origins with the work of von Mises in the area of probability and statistics (Knuth, 1997). While most experts feel all random probabilities are by nature actually pseudo-random in nature, a sub-field of statistical communication theory, also known as information theory, has developed a standard measure of randomness known as Kolmogorov randomness, also known as Martin-Lof randomness, which was developed in the 1960's (Knuth, 1997; Shannon, 1948; M. and Vitanyi, 1997). This sub-field of information theory is known as algorithmic information theory (Ge, 2005). What makes this measure of randomness, and non-randomness,so distinct is the notion of patterns and pattern less, sequences of 1's and 0's in a string of binary symbols (Martin-Lof, 1966). In other words, perceptual patterns as seen in a sequence of objects whichcan be defined as having similar sub-groupings within the body of the sequence that have a frequency, depending on the length of the string, of either regularity, non-randomness, orinfrequency, and randomness, within the sequence itself (Uspensky, 1992).

In examining the classification of a random and non-random set of 1's and 0's in two examples of a sequence of binary strings, the pattern verses patternless quantities can be examined.

Example 1 is as follows: [111000111000111] and example 2 is as follows: [110111001000011]. It is clear than example 1 is more patterned than example 2.Example 1 has a balanced sub-groups of three characters, either all 1's orall 0's, that have a perceptual regularity. Example2 is a classifier modeof a sequence of a random binary string in that the sub-groups, if grouped into like, or similar, characters, either 1's or a0's like in example 1, the frequency of the types of characters, either 1's or 0's, is different, seven variations of groups as opposed to the five variations in example 1, as are the sub-groups: [(11), (0), (111), (00), (1), (0000), and (11)] from example 2. While this would support the pattern verses pattern-less model proposed by Kolmogorov complexity, there is a strikingresult from these two examples1 and 2, in that the second, or random, example, Example 2, has a pattern within the sub-groups, that for all perceptual accounts, has distinct qualities that can be used to measure the nature of randomness on a sub-groupedlevel on examination of a binary string.

The author has done early work on coding each of the sub-groups and reducing them to a compressed state, and then decompressing them with no loss to either the amount of frequency or number of characters to a sequence of a binary string that would be considered random by Kolmogorov complexity (Tice). Now, while this simple program of compression and decompression by the author is for a future paper, the real interest of this bookis on thesub-groups as they stand without the notion of compression.

The very idea of the notion of a patterned or patternless quality as found in the measure of such aspects to the subgroupings of l's and 0's in a sequence of a binary string has the quality of being a bit vague, in that both example 1 and 2 are patterned, in which they have a frequency and similar character sub-groupings that have a known measure and quality and can be quantified in both examples. This is more than a question of semantics as the very nature of the measure of Kolmogorov complexity is the very fact that it has a perceptual "pattern" to measure the randomness of a sequence of a binary string. In reviewing the literature on the notions of patterns in Kolmogorov complexity/algorithmic information theory the real question arises, which patterns qualify for status as random, especially as a measure in a sequence of a binary string?

APPENDIX I

A RADIX 4 BASED SYSTEM FOR USE IN THEORETICAL GENETICS

ABSTRACT

The appendix will introduce the quaternary, or radix 4, based system for use as a fundamental standard beyond the traditional binary, or radix 2, based system in use today. A greater level of compression is noted in the radix 4 based systems when compared to the radix 2 base as applied to a model of informationtheory. The application of this compression algorithm to both deoxyribonucleic acid (DNA) and ribonucleic acid (RNA) sequences for compression will be reviewed in this appendix.

INTRODUCTION

A quaternary, or radix 4 based system, is defined as four separate characters, or symbols, that have no semantic meaning apart from not representing the other characters. This is the same notion Shannon gave to the binary based system upon its publication in 1948 (Shannon, 1948). This appendix will present research that shows the radix 4 based systems to have a compression value greater than the traditional radix 2 based system in use today (Tice, 2008). The compression algorithm will be used to compress DNA and RNA sequences. The work has applications in theoretical genetics and synthetic biology.

RANDOMNESS

The earliest definition for randomness in a string of 1's and 0's was defined by von Mises, but it was Martin-Lof's paper of 1966 that gave a measure to randomness by the patternlessness of a sequence of 1'sand 0's in a string that could be used to define a random binary sequence in a string (Kotz and Johnson, 1982; Martin-Lof, 1986). A non-random string will be able to compress, were as a random string of characters will not be able to compress. This is the classical measure for Kolmogorov complexity, also known as algorithmic information theory, of the randomness of a sequence found in a binary string.

COMPRESSION PROGRAM

The compression program to be used has been termed the modified symbolic space multiplier program as it simply notes the first character in a line of charactersin a binary sequence of a string and sub-groups them into common or like groups of similar characters, all l's grouped with 1's and all 0's grouped with 0's, in that string and is assigned a single character notation that represents the number found in that sub-group, so that it can be reduced, compressed, and decompressed, expanded, back to its original length and form (Tice). An underlined 1 or 0 is usually used to note the notation symbol for the placement and character type in previous applications of this program. The underlined initial character to be compressedwill be used for this appendix.

APPLICATION OF THEORY

The application of a quaternary, or radix 4 based system, to existing genetic marking and counting systems has many advantages. The first is the greater amount of compression from this base, as opposed to the standard binary based system in use today, and secondly, as a more utilizable system because of the four characters, or symbol, based system that provides for more variety to develop information applications.

DEOXYRIBONUCLEIC ACID (DNA)

The DNA is a linear polymer made up of specific repeating segments of phosphodiester bonds and is a carrier of genetic information (Lutter, 2007). There are four bases in DNA—adenine, thymine, guanine, and cytosine (Lutter).The use of a compression algorithm for sequences of DNA.

DEFINITIONS

A = Adenine
T = Thymine
G = Guanine
C = Cytosine

EXAMPLEA:

ATATGCGCTATACGCGTATATATA

The compression algorithm will use a specific focus on TA and GC DNA sequences in example A.

KEY CODE

TA = 4 characters

GC = 2 characters

COMPRESS EXAMPLE: A

ATATGCATATCGCGTA

The compressed DNA sequenceis 16 characters from the original non-compression total of 24.

The use of a four character system, a radix 4 base number system, which is composed of each character not representing the other characters, is ideal in DNA sequences composed of adenine, thymine, guanine, and cytosine.

EXAMPLE: D

TAGCTAGCTAGCTAGCTAGCTAGCTAGCTAGCTAGCTAGC

KEY CODE

TAGC = 10

COMPRESSION OF EXAMPLE D

TAGC

The compressed version of example D is 4 characters from the original non-compressed total of 40 characters.

RIBONUCLEIC ACID (RNA)

The RNA translates the genetic information found in DNA into proteins (Beyer, 2007). There are four bases that attach to each ribos (Beyer).

The use of a compression algorithm for sequences of RNA.

DEFINITIONS

A = Adenine
C = Cytosine
G = Guanine
U = Uracil

EXAMPLE B

AUAUCGCGAUAUCGCGUAUAUAUAGCGC
The compression algorithm will focus on specific RNA sequences.

KEY CODE

UA = 4 characters
GC = 2 characters
Compress Example #B
AUAUCGCGAUAUCGCG<u>UAGC</u>

The compressed RNA sequence is 20 characters in length from the original non-compression total character length of 28.

The use of a four character system, a radix 4 base number system, which is composed of each character not representing the other characters, is ideal in RNA sequences composed of adenine, cytosine, guanine, and uracil.

EXAMPLEC

UAGCUAGCUAGCUAGCUAGCUAGC
The use of a universal compression algorithm is as follows:
Key Code
UAGC = 6

COMPRESSION OF EXAMPLE C

UAGC

The compressed version of example C is 4 characters from the original non-compressed 24 character total length.

SUMMARY

The compression algorithm used for both DNA and RNA sequences has the power of both universal compression algorithm, all character length types, and a specific, or target, level of compression.

APPENDIX J

A COMPRESSION PROGRAM FOR CHEMICAL, BIOLOGICAL, ANDNANOTECHNOLOGIES

ABSTRACT

The appendix will introduce a compression algorithm that will use based number systems beyond the fundamental standard of the traditional binary, or radix 2, based system in use today. A greater level of compression is noted in these radix based number systems when compared to the radix 2 base as applied to a sequential string of various information. The application of this compression algorithm to both random and non-random sequences for compression will be reviewed in this paper. The natural sciences and engineering applications will be areas covered in this appendix.

INTRODUCTION

A binary, or radix 2 based, system is defined as two separate characters, or symbols, that have no semantic meaning apart from not representing the other character. This is the same notion Shannon gave to the binary based system upon its publicationin 1948 (Shannon, 1948). This appendix will present research that shows how various radix based number systems have a compression value greater than the traditional radix 2 based system as in use today (Tice, 2008). The compression algorithm will be used to compress various random and non-random sequences. The work has applications in theoretical and applied natural sciences and engineering.

RANDOMNESS

The earliest definition for randomness in a string of 1's and 0's was defined by von Mises, but it was Martin-Lof's paper of 1966 that gave a measure to randomness by the patternlessness of a sequence of 1's and 0's in a string that could be used to define a random binary sequence in a string (Kotz, S. and Johnson, 1982; Martin-Lof, 1966).

A non-random string will be able to compress, were as a random string of characters will not be able to compress. This is the classical measure for Kolmogorov complexity, also known as algorithmic information theory, of the randomness of a sequence found in a binary string.

COMPRESSIONPROGRAM

The compression program to be used has been termed the modified symbolic space multiplier program as it simply notes the first character in a line of characters in a binary sequence of a string and sub-groups them into common or like groups of similar characters, all 1's grouped with 1'sand all 0's grouped with 0's, in that string and is assigned a single character notation that represents the number found in that sub-group, so that it can be reduced, compressed, and decompressed, expanded,back to its original length and form (Tice). An underlined 1 or 0 is usually used to note the notation symbol for the placement and character type in previous applications of this program. The underlined initial character to be compressed will be used for this appendix.

APPLICATIONOF THEORY

The compression algorithm will be used for the following radix based number systems: Radix 6, Radix 8, Radix 10, Radix 12, and Radix 16. These are traditional radix base numbers from the field of computer science and have strong applicationsto other fields of science and engineering due to the parsimonious nature of these low digit radix base number systems (Richards, 1955). The compression algorithm in this appendixcan be both a "universal" compression engine in that all members of a sequence, either random or non-random, can be compressed or a "specific" compression engine that compressesonly specific types of sub-groups within a random or non-random string of a sequence.

The compression algorithm will be defined by the following properties:

1. Starting at the far left of the string, the beginning, and moving to the right, towards the end of the string.
2. Each sub-group of common characters, including singular characters, will be grouped into common sub-groups and marked accordingly.
3. The notation for marking each sub-group will be underling the initial character of that common sub-group. The remaining common characters in that marked sub-group will be removed. This results in a compressed sequential string.
4. Decompression of the compressed string is the reverse process with complete position and character count to the original pre-compressed sequential string.
5. This will be the same processes for both random and non-random sequential strings.

CHEMISTRY

Chemistry is the science of the structure, the properties and the composition of matter and its changes (Moore, 1993).

POLYMER

A polymer is macromolecule, large molecule, and made up of repeating structural segments usually connected by covalent chemical bonds (Wikipedia "*Polymer*", 2010).

COPOLYMER

A copolymer, also known as a heteropolymer, is a polymer derived from two or more monomers(Wikipedia "*Copolymer*, 2010).

Types of Copolymers are:

1. Alternating Copolymers Regular alternating A and B units.
2. Periodic Copolymers: A and B units arranged in a repeating sequence.
3. Statistical Copolymers Random sequences.
4. Block Copolymers Made up of two or more homopolymer subunits joined by covalent bonds.
5. Stereoblock Copolymer A structure formed from a monomer.

An example ofthe use of a compression algorithm on copolymersis as follows:

1. Alternating Copolymers Alternating copolymers using a radix 2 base number system.

Unit A = 0
Unit B = 1

EXAMPLE 1:

01010101010101
Compression of Example1

KEY CODE

0 = 7 characters
1 = 7 characters

EXAMPLE 1COMPRESSED

01

The compressed state of example 1 is a 2 character length from the original non-compression state total of 17 characters in length.

Periodic Copolymers: Periodic copolymers using a radix 16 base number system.

Unit A = abcdefghijklmnop
Unit B = 123456789@#$%"&*

EXAMPLE 2

abcdefghijklmnop123456789@#$%^&*123456789@#$%^&*abcdefghijklmno pl23456789@#$%^&*

Compression of Example 2

KEY CODE

abcdefghijklmnop = 16 characters
123456789@#$%^&* = 16 characters

EXAMPLE 2 COMPRESSED

a1la1

The compressed state of example2 is 5 characters from the original non-compression state total of a 80 character length.

Statistical Polymers: Random copolymer using a radix 8 base number system.

Unit A= 12345678
Unit B = abcdefgh

EXAMPLE 3

1 2 3 4 5 6 7 8 a b c d e f g h a b c d e f g h 1 2 3 4 5 6 7 8 1 2 3 4 5 6 7 8 abcdefgh123456781234567812345678

KEY CODE

12345678 = 8 characters
abcdefgh = 8 characters

COMPRESSIONOF EXAMPLE 3

1a11alll

The compressed state of example 3 is 8 from the original non-compression state total of a 64 character length.

Block Copolymers: Block copolymer using a radix 12 base number system.

Unit A = abcdefghijkl
Unit B = 123456789@#$ Example#$

1 2 3 4 5 6 7 8 9 @ # $ 1 2 3 4 5 6 7 8 9 @ # $ 1 2 3 4 5 6 7 8 9 @#$abcdefghijklabcdefghijklabcdefghijkl

KEY CODE

abcdefghijk = 12 characters
123456789@#$ = 12 characters

COMPRESSIONOF EXAMPLE 4

111aaa

The compressed state of Example #4 is 6 characters from the original non-compression state of 58 character length.

Stereoblock Copolymer: Stereo block copolymer using a radix 10 base number system.

Unit A = abcdefghij

Unit B = 123456789@

Note: The symbol [I] represents a special structure defining each block.

EXAMPLE 5

abcdefghijabcdefghijabcdefghijabcdefghijabcdefghijabcedfghij

I I

123456789@123456789@ 123456789@123456789@

KEY CODE

abcedfghij = 10 characters

123456789@ = 10 characters

COMPRESSIONOF EXAMPLE 5

aaaaaa

1111

The compressed state of example 5 is 10 characters from the original non-compression total of 100 characters in length.

BIOLOGY

Biology is the study of nature and as such is a part of the systematic atomistic axiomization of processes found within living things. These natural grammars or laws have mathematical corollates that parallel process found in the physical and engineering disciplines. The use of a compression algorithm of a sequential string is a natural development of such a process as can be seen in the compression of both deoxyribonucleic acid (DNA) and ribonucleic acid (RNA) genetic codes.

DEOXYRIBONUCLEIC ACID (DNA)

The DNA is a linear polymer made up of specific repeating segments of phosphodiester bonds and is a carrier of genetic information (Lutter, 2007). There are four bases in DNA—adenine, thymine, guanine, and cytosine (Lutter). The use of a compression algorithm for sequences of DNA.

DEFINITIONS:

A = Adenine

T = Thymine

G = Guanine

C = Cytosine

EXAMPLE A

ATATGCGCATATCGCGTATATATATATA

The compression algorithm will use a specific focus on TA and GC DNA sequences in example A.

KEY CODE

TA = 6 characters
GC = 2 characters

COMPRESS EXAMPLE A

ATATGCATATCGCGTA

The compressed DNA sequenceis 16 characters from the original non compression total of a 28 character length.

RIBONUCLEIC ACID (RNA)

RNA translates the genetic information found in DNA into proteins (Beyer L and Gray, 2007). There are four bases that attached to each ribose (Beyer).

DEFINITIONS:

A = Adenine
C = Cytosine
G = Guanine
U = Uracil

EXAMPLE B

AUAUCGCGAUAUCGCGUAUAUAUAUAUAGCGC

The compression algorithm will focus on specific RNA sequences.

KEY CODE

UA = 6 characters
GC = 2 characters

COMPRESS EXAMPLE B

AUAUCGCGAUAUCGCGUAGC

The compressed RNA sequence is 20 characters in length from the original non-compression total character length of 32.

NANOTECHNOLOGY

The development and discovery of nanometer scale structures, ranging from 1 to 100 nanometers, to transform matter, energy, and information on a molecular level of technology (Drexler, 2007).

SYNTHETIC BIOLOGY

Within the field of synthetic biology is the development of synthetic genomics that uses aspects of genetic modification on pre-existing life forms to produce a product or desired behavior in the life form create (Wikipedia "*Synthetic genomics*", 2010).

The following is a DNA sequence of real and "made up" synthetic sequences.

DEFINITIONS

A = Adenine
T = Thymine
G = Guanine
C = Cytosine
W = *Watson
K = *Crick
*Note: Made up synthetic DNA.

EXAMPLE C

TATAGCGCWKWKATATCGCGKWKWKWKWKWKW

KEY CODE

AT = 2 characters
CG = 2 characters
KW = 6 characters

COMPRESSEDEXAMPLE C

TATAGCGCWKWKATCGKW

The compressed synthetic DNA sequence is 18 characters from the original non-compression character total of 32.

SUMMARY

The appendix has addressed the use of a compression algorithm for use in various radix based number systems in the fields of chemistry, biology, and nanotechnology. The compression algorithm in both the universal and specific format have successfully reduced long and short sequences of strings to very compressed states and function well in both random and non-random sequential strings.

APPENDIX K

STATISTICAL PHYSICS AND THE FUNDAMENTALS OF MINIMUM DESCRIPTION LENGTH AND MINIMUM MESSAGE LENGTH

ABSTRACT

This monograph is the first account of the use of a "summing engine" an algorithm for counting and unifying common, or liked natured, characters in a binary sequential string and combining them into a sub-group of common character types into a compressed collective.

The ability to compress a random binary sequential string is a novel feature of this "summing engine" algorithm and is examined in light of both minimum description length and its progenitor minimum message length that use data compression as a parameter for defining "good" data for evaluating a model for measurement.

A new paradigm for both minimum message length and minimum description length results from this study.

INTRODUCTION

The monograph will address the "compressibility" of a traditional random binary segmental string verses a "summing engine" binary sequential string. The model used for both compression systems are the minimum message length (MML), and minimum description length (MDL), with the result being a paradigm shift of both the MML and MDL systems at the fundamental level.

The MML was first developed by Wallace and Boulton (1968). The bases of MML is very similar to MDL except the MML is a fully subject Bayesian model (Wikipedia, 2011).

The MDL was first developed by Rissanen in 1978 (Rissanen, 1978). The base of the MDL is that regularity in a specific set of data can be used to compress the data into a sequence shorter than the original length of the originating sequence.

MINIMUM MESSAGE LENGTH

Minimum message length (MML) was the early progenitor of minimal description length (MDL) and was first published by Wallace and Boulton in 1968. The primary difference between MML and MDL, the MDL is that the MML is a fully subjective Bayesian model in that it is of a 'a prior' distribution (Wikipedia, 2012).

In the MML model all the parameters are encoded in the first part of a two-part code so all the parameters are learned (Wikipedia, 2012).

MINIMUM DESCRIPTION LENGTH

Minimum description length (MDL) was developed in 1978 by Jorma Rissanen as a method by which the "best hypothesis for a given set of data is the one that leads to the best compression of the data" (Wikipedia, 2012). The MDL has to bypass two fundamentals of Kolmogorov complexity in that Kolmogorov complexity is incomputable and uses "what" computer language is used (Wikipedia, 2012). The MDL restricts the codes allowable so that it does become computable to find the shortest code length available and that the code being "reasonably" efficient (Wikipedia, 2012).

The MDL Principle is as follows:

The best theory minimizes the bit sum of the length of the description of the theory and the length in bits of the data when encode by the help of the theory (Li and Vitany, 1997). The MDL tries to balance the regularity and randomness in the data using the best model, the one that uses regularity in the data to compress (Li and Vitanyi, 1997).

THE GRAMMAR OF FORM

Martin-Lof (1966) has developed a form of algorithmic complexity based on patterns within sequential strings of binary data of both a random and non-random manner (Martin-Lof, 1966). A more regular pattern of a binary sequential string would look like this:

[1010101010]

This type of regularity has regularity, a balance, of form of both [0's] and [1's] that marks it as a non-random sequential string (Martin-Lof, 1966). A random sequential string would look like the following:

[0100011000]

These patterns represent the patterns of the grammar of statistical randomness and are the tradition measure of "the qualities and quantities" of the notion of statistical randomness in a sequential string.

COMPRESSION ENGINE

A "compression engine" is, in essence, an algorithm for common types of character's, either all [0's] or [l's], in a binary sequential string to compress into like-natured characters and then bede-compressed when desired (Tice, 2009). Much of the "ornamental", visual markers, found in my first large scale address of a "summing engine" is removed from this publication to save time and extraneous notation (Tice, 2009).

The "summing engine" is a systematic process of the sequential addition of common or liked natured characters. In this case the binary [l's] and [0's] of minimally weighted semantic values that have the [l's] being the opposite value of the [0's].

A NEW PARADIGM

If a "summing engine" is used on a binary sequential string the following three results will occur:

1. The pattern of the binary sequential string will be of a regular pattern or non-random distribution.
2. The pattern of the binary sequential string will be of a non-regular pattern or of a random distribution.
3. The pattern of the binary sequential string will be of both a regular and a random distribution.

Along with the three types of binary sequential string types, the fundamental properties of "a priori" and "a posteriori" mark the pre-algorithm and the post-algorithm models.

A traditionally compressed binary sequential string would be non-random as follows:

[1010101010]

A traditional binary sequential string would not compress as follows:

[0111100011]

If a "summing engine" is used as the algorithm both non-random and random binary sequential strings would both compress as follows:

NON-RANDOM COMPRESSION:

[1010101010]

Compressed:

[10] Five times

RANDOM COMPRESSION:

[1000110000] with [1×1] [0×3] [1×2] [0×4] or [1010]

If the "summing engine" is used along with the traditional notions of Kolmoqorov complexity a great deal of change is noted for the "summing engines" ability to compress a random binary sequential string as to the traditional Kolmoqorov complexity random model.

Minimal description length avoids assumptions about data generating procedures were as minimum message length "represents a Bayesian framework" (Wikipedia, 2012). It has been noted in the literature that some researchers feel that minimum description length is equivalent to Bayesian inference but is renounced by Rissanen as being data which does not reflect the "true' nature of whether the data as collected or the data to 'reality' (Wikipedia, 2012).

Because two forms of randomness exist traditional, unable to compress randomness, and the compressible type of a "summing engine", able to compress a random sequential string, the fundamentals of both the minimum message length and more developed model of minimum description length are divergent at the point of what makes up a "random binary sequential string and compression".

Modern minimum description length theory is based on the principle of compression, as is traditionally known; Kolmogorov complexity, and has no development using a random binary sequential string using a "summing engine" (Wikipedia, 2012 and Li and Vitanyi, 1997).

CONCLUSION

Using both the minimal message length model and minimal description length model as "tests" of randomness found in binary sequential strings against both the traditional Kolmogorov complexity notion of randomness and the author's "summing engine" form of randomness in binary sequential strings result in two different sets of measures.

The resulting changes to both the minimal message length and minimal description length models is at the fundamental level of statistical physics and adds a new chapter to the study of algorithmic complexity.

SUMMARY

This monograph is the first account of a major fundamental change to both minimum message length and minimum description length. The application of a "summing engine" to make a random binary sequential string to compress has foundational effects to the traditional notions about compression as a fundamental level of the parameters for the measure of such compression found in both the minimum message length and minimum description length models.

NOTES

Rissanen notes his, with Barron and Yu, that Wallace and Boulton's seminal paper (1968) to the field of minimal description length as being the "crudest". Wallace and Boulton's paper of 1968 seems more "rudimentary" than "crude" and is, in some respects, a different set of measures than Rissanen's work (Barron, Rissanen, and Yu, 1978 and Wallace and Boulton, 1968).

APPENDIX L

THE USE OF SIGNAL FLOW DIAGRAMS IN PHARMACOLOGY

The use of signal flow diagramsare common in fields such as engineering. A practical use can be made in the field of pharmacology for similar reasons-ease of use. These graphs can be used to solve complex linear, multiloop systems in less time than either block diagrams or equations (Macmillan, Higgins, and Naslin, 1964). A signal flow graph is the topological representation of a set of linear equations as represented by the following equation:

Equation 1: Yi = L. aij Xj,i = 1,..., n

Branches and nodes are used to represent a set of equations in a signal flow graph. Each node represents avariable in the system, like node i represents variable y in equation 1. Branches represent the different variables such as branch ij relates variable yi to yj where the branch originates at node i and terminates at node j in equation 1 (Shinners, 1964). The following set of linear equations are represented in the signal flow graph in Figure 1 (Shinners, 1964).

y_2 = ay, + by2 +by3
y_3 = dy2
y_4 = ey1 + fy3
y_5= + gy3 + hy4

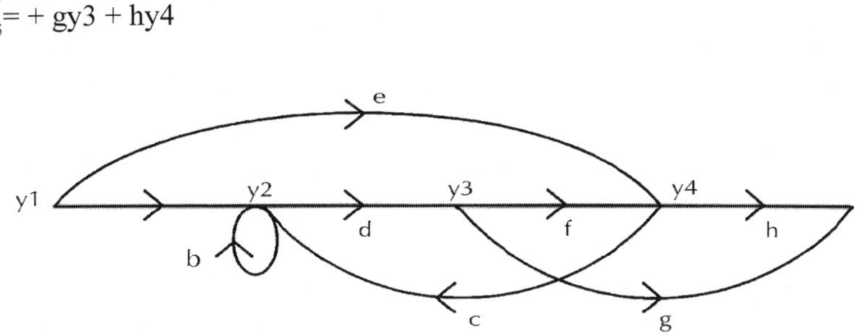

FIGURE 1 Signal flow graph.

By using a signal flow diagram to represent time of drug delivery, behavioral response to the drug, and drugdosage, the cumulative sequence of events is visually represented for ease of interpretation of data.

This type of graphic representation of complex linear equations makes interpretation of drug analysis that much more efficient and effective for evaluation.

APPENDIX M

SIGNAL FLOW DIAGRAMS VERSES BLOCK DIAGRAMS

The strongest points in using a signal flow diagram over block diagrams is that they are more general in use and are a more pronounced rationalization afforded by diagrammatic simplification (Macmillian, Higgins, and Naslin, 1964).

METHODS

A block diagram is a graphical representation of interconnected elements which form a system that differ only in their dynamic properties (Solodov, 1966). A signal flow diagram represents a set of equations by means of branches and nodes. The use of the signal flow diagram permits the solution practically by visual inspection (Shinners' 1964). A model of a block diagram will becompared to a model of a signal flow diagram for a comparative and contrastive analysis of the two graphic systems.

EXAMPLES

The following is a model of a standard single-loopfeedback system using a block diagram in Figure l (Eveleigh, 1960).

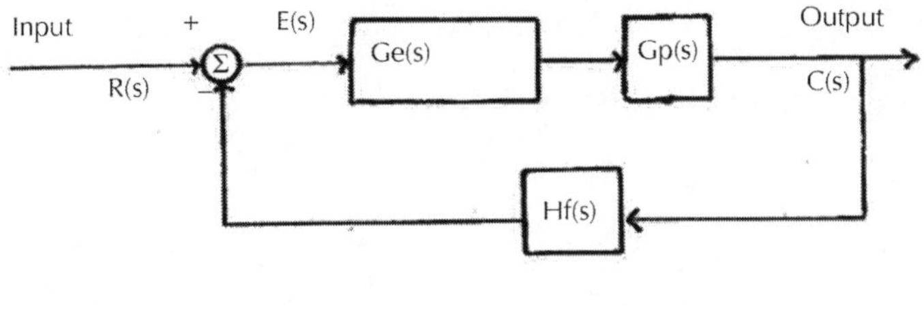

FIGURE 1 Block diagram of standard single-loop feedback system.

The next model is a signal flow diagram for the following algebraic equation system, Table 1, and represented by a signal flow diagram in Figure 2 (Macmillian, 1964).

TABLE 1

x = x
x = t x + t x
x = t x + t x + t x

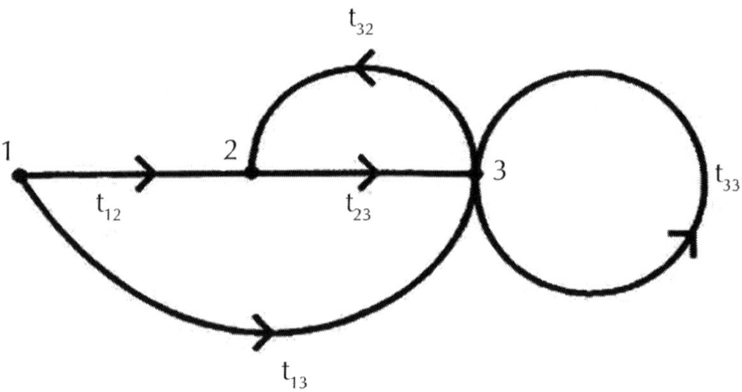

FIGURE 2 Signal flow diagram of algebraic equation system.

DISCUSSION

The strengths of the signal flow diagram over the block diagram are very pronounced on all levels of evaluationand merit a wider application of such a graphing system to other information display categories. These qualities of simplicity, generality of use, and clarity of realization makes the signal flow diagram amore efficient method than block diagrams on solving a systems problem.

RESULTS

In examining the block diagram in Figure 1 and the signal flow diagram in Figure 2 a comparison between the two systems can be made on a point by point analysis.

1. The signal flow diagram is a simplified graphic representation of mathematical, numerical, and word models and that these models are best expressed by the signal flow diagram.

2. The nodes and branches of the signal flow diagram can be a symbol of all types of data and information and the branches' arrows can represent loops, increases and decreases in relation to the information being represented and the nodes denoting a hierarchy of the information being represented.
3. Time and space are saved by the use of signal flow diagrams and this can be important when labor, cost, and efficiency factors are involved.
4. The complexity of the information is made simpler and more visually clear by the use of signal flow diagrams and that this simplicity is inherent in such a graphical method.
5. The signal flow diagram is superior to formal or blocks diagrams as block diagrams are inherently weak in the area of simplification and ease of use and are also time and space sensitive.
6. Both equations and raw data are inferior to signal flow diagrams in that equations are long, complex and time consuming and raw data is marginal at imparting
7. The signal flow diagram is superior to tables, charts andflow charts in that it more readily accepts large quantities of information and represents them in the most accurate and simplistic manner, something that tables, charts and flow charts perform with limited success.
8. The information saturation point of signal flow diagrams is higher than other forms of information representation.
9. Overall simplicity of conception, use and understanding is the main point of interest and support for the signal flow diagram.
10. Accuracy of the signal flow diagram is in the simplicityof its use. Discussion

APPENDIX N

The data and information in Model A is taken from Lynn E. Spiter's *Delayed Hypersensitivity Skin Testing* from Rose and Friedman's *Manual of Clinical Immunology* (Washington: American Society for Microbiology, 1980).

Skin testing is the most important clinical assessment of the status of cellular immune responses in patients (Rose and Friedman, 1980). In essence, a delayed hypersensitivity reaction is if a red bump develops at the site of injection of a test antigen, indicating that the afferent, central, and efferent limbs of the immune system response is intact and that the patient's ability to start a nonspecific inflammatory response is intact (Rose and Friedman, 1980). There are four clinical indications that skin tests are used to assess.

1. To assess whether there is diminished delayed hypersensitivity or anergy in selected patients.
2. To assess the results of immune therapy.
3. To follow the course of the disease process.
4. As an aid to diagnosis infectious diseases.

PROCEDURES

The test procedures are as follows:

1. Battery of six skin test antigens.
2. Repeat the test in higher antigen concentrations when the tests are negative withthe intermediate strength.
3. Observe and record the results in millimeters of erythema and induration at 24 hrs and 48 hrs intervals (Rose and Friedman, 1980).

The six antigens are to be injected into a marked region, indelible pencil, on the patient, usually the forearms, in a subcutaneous fashion leaving a 5 cm diameter bleb. Recording the results after 24 hrs (Rose and Friedman,1980).

INTERPRETATION

A positive response is determined when a 5 mm or more of induration is found at the test site 48 hrs after the injection of the test antigen.

ANTIGENS

The six antigens are as follows:

1. Candidin
2. Mixed respiratory vaccine
3. PPD
4. SK-SD
5. Staphage Lysate
6. Trichoplytin

Note: Both Coccidioid in and Mumps were used in the test in gas they are also clinical indicators of possible energy.

APPENDIX O

The data and information for the "Triple-Test Plan for Serologic Diagnosis of Syphilis" is from Levinson's and MacFate's *Clinical Laboratory Diagnosis* (Philadelphia: Lea and Febiger, 1969).

There are many serologic tests that canbe used to detect the etiologic agent for syphilis, the *Treponema pallidu*m, and its antibodies, in serums and cerebrospinal fluids. The tests may be treponemal or nontreponemal (Levinson and MacFate, 1969). Nontreponemal tests include precipitation and flocculation methods, such as Veneral Disease Research Laboratory (VDRL) test and the Kahn test, and those complement-fixation tests which use antigens derived from animal tissue extracts such as the Kolmer test (Levison and MacFate, 1969).

Treponemal tests include agglutination and treponemal immobilization methods, such as the *Treponema Pallidum* Immobilization (TPI) test, and those complement-fixation tests which use antigens extracted from either virulent or saprophytic strains of *Treponema pallidum*, such as the Reiter Protein Complement-Fixation test (Levison and MacFate, 1969). Most positive serologic tests obtained with nontreponemal antigens are due to syphilis (Levisnson and MacFate, 1969).

A biologic false-positive reaction may be due to the presence of antibody like substances similar to the antibodies produced in syphilis. To overcome these false reactions, tests were developed using the *Treponema pallidum* itself or an antigen derived from the organism. These tests are specific and biologically false-positive reactors give negative results (Levinsonand MacFate,1969).

APPENDIX P

The data and information for cyclosporine is taken from drug facts and comparisons (St. Louis: Facts and Comparisons, 1995), Harkness's *Drug Interactions Guidebook* (Englewood Cliffs: Prentice Hall, 1991), and Rybacki and Long's *The Essential Guide to Prescription Drugs* (New York: Harper Perennial, 1996).

Cyclosporine is a cyclic polypeptideimmunosup pressant. The available dosage forms and strength are as follows:

1. Capsules, soft gelatin-25 mg, 100 mg.
2. Injection, intravenous-50 mg per ml.
3. Oral solution-100 mg per ml.

Usual dosage range is 15 mg per day taken 4–12 hr before transplant surgery. Reduced to 5–10 mg per day two weeks after surgery (Rybacki and Long, 1996).

Cyclosporine blood levels should be closely monitored along with the serum creatinine. Regulate, that is lower dosage level, dose of cyclosporine as needed (Harkness, 1991).

Distribution of cyclosporine in the body is as follows:

1. 33–47% is in plasma.
2. 4–9% in lymphocytes.
3. 5–12% in granulocytes.
4. 41–58% in erythrocytes.

Note: Blood level monitoring of cyclosporine may be useful in patient management. While no fixed relationships have been established, 24 hr trough values of 250–800 ng/ml (wholeblood, RIA) or 50–300 mg/ml (plasma, RIA) appear to minimize side effects and rejection events (Facts and Comparisons, 1995).

A LIST OF THE EDITOR'S PAPERS ON SIGNAL FLOW DIAGRAMS

PROFESSIONAL PAPERS

"A Theoretical Model of Feedback in Pharmacology Using the Signal Flow Diagram". Ph.D. Dissertation (1996). Published by U.M.I., Ann Arbor, Michigan U.S.A.

"The use of signal flow diagrams in molecular biology." Essay that was awarded a "Certificate of Merit" from the Pharmacia Biotech & Science Prize for Young Scientists 1997.

"Signal Flow diagrams and biotechnology." Poster presented at the Annual Spring Meeting of the NCASM, Northern California Chapter of the American Society for Microbiology, Apri115.1997, at Santa Clara University, Santa Clara, California U.S.A.

"The use of signal flow diagrams in chemical analysis." Poster presented at the 15th Annual American Peptide Symposium, June 17, 1997, Nashville, Tennessee.

"Chromatic aspects of the signal flow diagram." Poster presented at the 78th Annual Meeting of the Pacific Division of the Ameriean Association for the Advancement of Science-AAAS, June 23–24, 1997, Corvallis, Oregon, U.S.A.

"The multicolored arrows: Chromatic aspects of the signal flow diagram." Poster presented at the214th American Chemical Society–ACS, September 7–11, 1997, Las Vegas, Nevada U.S.A.

"Signal flow diagrams as predictors of reliability in systems." Poster presented at the 79th Annual Meeting of the Pacific Division of the American Association of the Advancement of Science–AAAS, June 28–July 2, 1998, Logan, Utah U.S.A.

"Feedback systems for nontraditional medicines: A case for the signal flow diagram." Journal of Pharmaceutical Sciences. Volume 87, Number 11, pp.1282–1285. November 1998

REFERENCES

1. Adolph, E. F. *The Development of Homeostasis*. Academic Press, London (1960).
2. Ahrendt, W. R. *Servomechanism Practice*. McGraw-Hill Book Company Inc., New York (1954).
3. Alagoz, B. B. *Effects of sequence partitioning on compression rates*. (November 3, 2010) pp. 1–6. Website: http://arxiv.org/abs/1011.0338
4. Alexander, S. B. Optical *Communication Receiver Design*, SPIE Optical Engineering Press, London, UK, Chapter 6, pp. 173–201 (1997).
5. Barron, A., Rissanen, J., and Yu, B. The Minimum Description Length Principle in Coding and modelling. *IEEE Transactions on Information Theory*, **44**(6), 1–17 (October 1998).
6. Barry, J. R. *Wireless Infrared Communications*, Kluwer Academic Pub., Boston, Mass., Chapter 3, pp. 49–52, 1994.
7. Battelle, J. *The Search Portfolio*, New York (2005).
8. Bellazzi, R. Drug delivery optimization through Bayesian networks, an application to erythropoietin therapy in uremic anemia. *In Computers and Biomedical Research*, **26**(3), 274–93 (June, 1993).
9. Berners-Lee, T. *Weaving the Web, Harper*, San Francisco (1999).
10. Beyer, A. L. and Gray, M. W. *Ribosomes*. In McGraw-Hill Encyclopedia of science & technology. McGraw-Hill Publishers, New York, pp. 542–546 (2007).
11. Beyer, abide. p. 5.
12. Beyer, abide. p. 542.
13. Blackman, R. B. Effect of Feedback on Impedance, *Bell System Tech. Journal*, **22**, 269–277 (October, 1943).
14. Bobrow, L. S. *Elementary Linear Circuit Analysis*, 2nd Ed., HRW Inc., New York (1987).
15. Boehlert, S. Explaining scienceto power: make it simple, make it pay. *Science*, **314**, 1228–1229 (November 24, 2006).
16. Bornholdt, S. Less is more in modeling large genetic networks. *Science*, **310**, 449–451 (October 21, 2005).
17. Bowen, J. L. and Schultheises, P. M. *Introduction to the Design of Servomechanisms*. John Wiley and Sons Inc., New York (1961).
18. Brandman, O, Ferrell, J. E., Li, R., and Meyer, T. Interlinked fast and slow positive feedback loops drive reliable cell decisions. *Science*, **310**, 496–498 (October 21, 2006) (2005).
19. Brass, E., Hilleringmann, U., and Schumacher, K. System Integration of Optical Devices and Analog CMOS Amplifiers. IEEE J Solid-State Circuits, **29**(8), 1006–1010 (August 1994).
20. Brecher, J. Graphical Representation of Stereo-chemical Configuration. *Pure and Applied Chemistry*, **78**(10), 1897–1970 (2006).
21. Breeding, K. J. *Digital Design Fundamentals.* Prentice Hall, Englewood Cliffs (1992).

22. Brookhaven National Laboratory. Researchers find surprising similarities between genetic and computer codes. *Phys.org.*, pp. 1–3, (March 29, 2013). Http://phys.org/news/2013-03-similarities-genetic-codes-html
23. Bulgakov, A. A. *Energetic Processes in Follow up Electrical Control Systems*. The Macmillan Company, New York (1965).
24. Bulliet, L. J. *Servomechanisms*. Addison-Wesley Publishing Company, Menlo Park (1967).
25. Bullock, T. H., Bennett, M. V. l., Johnston, D., Josephson, R., Marder, E, and Fields, R. D.
26. Busacker, R. G. and Saaty, T. L. *Finite graphs and networks: An introduction with applications*. McGraw-Hill Book Company, New York (1965).
27. Cannon, W. *the wisdom of the body*. W.W. Norton & Company Inc., New York (1932).
28. Carnegie Mellon University. *Grammar undercuts security of long computer passwords. Science Daily*. pp. 1–3 (January 27, 2013). Website: http://www.sciencedaily.com/releases/2013/01/130124123549.html
29. Carroll. J. B. *Language, Thought, and reality*. Cambridge: The MIT Press (1956).
30. Cartage.org. *The importance of structural formulas*. pp. 1–3 (March 16, 2013). Website: http://www.cartage.oeg.1b/en/themes/Sciences/Che,istry/Organi.
31. Caruthers, F. P. and Levenstein, H. *Adaptive Control Systems*. A Pergamon Press Book, New York (1963).
32. Chang, S. L. *Synthesis of Optimum Control Systems*. McGraw-Hill Book Company Inc., New York (1961).
33. Chem.qmul. 2-Carb-3 and 2-Carb-4. pp. 1–5 (March 16, 2013). Website: http://www.chem.qmul.ac.uk/iupac/2carb/033n04.html.
34. Chen, W. K. *Active Network Analysis*, World Scientific, Singapore (1991).
35. Chen, W. K. *Graph Theory and Its Engineering Applications*, World Scientific, Singapore (1997).
36. Cherniavsky, N. and Ladner, R. *Grammar-based compression of DNA sequences*. UW CSE Technical Report 2007-05-02. pp. 1–21 (May 28, 2004).
37. Cho, A. Effect that revolutionized hard drives nets a nobel. *Science*, **318**, 179 (October 12, 2007).
38. Chubb, B. A. *Modem Analytical Design of Instrument Servomechanisms*. AddisonWesley Publishing Company, Palo Alto (1967).
39. Clark, E. M. A new design for an audio dosimeter. *In the American Industrial Hygiene Association Journal*, **41**(10), 700–3 (October, 1980).
40. Coco, C. A more simple PROM programmer. *In Elettronica Oggi*, **2**(109–110), 112, 114–15 (February, 1979).
41. Condon, A, Harel, D., Kok, J. N., Salomaa, A., and Winfree, E. *Algorithmic Bioprocesses*, Berlin: Springer Publishing (2010).
42. Cronin, K. M., Lane, G. H., and Peirce, A. G. *Flow Charts: Clinical Decision Making in Nursing*. J. B. Lippincott, Philadelphia (1983).
43. Danilov, V. A. Television microscopes for studying biological microscopic material. *In Biomedical Engineering*. **18**(4), 125–130 (July–August, 1984).
44. Davies, F. *Impact of information technology touted*. Silicon Valley.com, p. 1. (Wednesday March 14, 2007)
45. Davis, A. M. Some Fundamental Topics in Introductory Circuit Analysis: Acritique, *IEEE Trans on Education*, **43**(3), 330–335 (August, 2000).
46. De Callatay, A. M. *Natural and Artificial Intelligence*. North Holland, New York (1992).
47. Del Toro, V. and Parker, S. R. *Principles of Control Systems Engineering*. McGraw-Hill Book Company Inc., New York (1960).

48. Derusso, P. M., Roy, R. J., and Close, C. M. *State Variables for Engineers*. John Wiley& Sons, New York (1965).
49. DeVries, S. H. and Baylor, D. A. An alternative pathway for signal flow from rod photoreceptors to ganglion cells in mammalian retina. *In the Proceedings of the National Academy of Sciences*, **92**(23), 10658–62 (November 1, 1995).
50. Doebelin, E. O. *Dynamic Analysis and Feedback Control*. McGraw-Hill Book Company Inc., New York (1962).
51. Drexler, K. E. *Nanotechnology*. In McGraw-Hill Encyclopedia of science & technology. McGraw-Hill Publishers, New York, pp. 604–607 (2007).
52. Dubourg, O., Chikli, F., and Delorme, G. Contrast echocardiography in clinical practice. *In the Journal d'Echographie et de Medecine par Ultrasons*. **16**(4), 143–152 (1995).
53. Durman, P. *Man who took google global*. Times on Line (The Sunday Times), pp. 1–4, (Sunday, May 20, 2007).
54. Eckschlager, K. and Danzer, K. *Information Theory in Analytical Chemistry*. New York: John Wiley & Sons (1994).
55. Eckschlager, K. and Stepanek, V. *Information Theory as Applied to Chemical Analysis*. New York: John Wiley & Sons (1979).
56. EMBL. *Researchers make DNA data storage a reality: every film and tv program ever created-in a teacup* (2013).
57. ESRC. How can we still raed words when the lettres are jmbbuled up?. *ScienceDaily*, (March 17, 2013). Website: http://www.sciencedaily.com/releases/2013/03/130315074613.htm
58. Evans, S. C. and Bush, S. F. *Symbol compression ratio for string compression and estimation of Kolmogorov complexity*. CiteSeerX. p. 1, (September 3, 2010) (2001). Website: http://citeseerx.ist.psu.edu/viewdoc/summary?doi=10.1.21.8207
59. Eveleigh, V. W. *Adaptive Control and Optimization Techniques*. McGraw-Hill Book Company, New York (1960).
60. Facts and Comparisons. *Drug Facts and Comparisons*. Facts and Comparisons, St. Louis (1995).
61. Faudree, R. *Graph Theory*. R. A. Meyers (Ed.) *Encyclopedia of Physical Science and Tachnolo9:i£•* Academic Press, New York, pp. 308–325 (1987).
62. Feil, H. J. Measuring, controlling and doserating with miniture oval-gear counters. In Und-oder Nor & Steuerungstechnik. **1–2**, 32–3 (1979).
63. Fukuyama, T., Yamaoka, K., Ohata, Y., and Nakagawa, T. A new analysis method for disposition kinetics of enterohepatic circulation of diclofenac in rats. *In Drug Metabolism Dispositions*. **22**(3), 479–485 (1994).
64. Gates, B., Myhrvold, N., and Rinearson, P. *The Road Ahead*, Viking, New York, p. 30 (1995).
65. Ge, M. *The New Encyclopedia Britannica*, Encyclopedia Britannica, Chicago, p.637 (2005).
66. Gille, J. C., Delegrin, M. J., and Deculine, P. *Feedback Control Systems*. McGraw-Hill-Book Company Inc., New York (1959).
67. Guenther, B., Morgado, E., and Penna, M. Equifinality on the circulatory system of some mammals. *In Pflug Archiv European Journal of Physiology*. **348**(4), 343–352 (1974).
68. H. Kresse! (Ed.) *Semiconductor Devices for Optical Communications*, 2nd Ed., Springer-Verlag, Berlin, Germany (1982).
69. Bode, H. W. *Network Analysis and Feedback Amplifier Design*, Van Nostrand, New York, (1945).
70. Hadley, W. A. and Longobardo, G. *Automatic Process Control*. Addison-Wesley Publishing Company Inc., Palo Alto (1963).

71. Hardie, A. M. *The Elements of Feedback and Control.* Oxford University Press, New York (1964).
72. Harkness, R. *Drug Interactions Guide book.* Prentice Hall, Englewood Cliffs (1991).
73. Haykin, S. S. *Active Network Theory*, Addison-Wesley, Reading, Chapter 10, Massachusetts (1970).
74. Heister, H. New developments in x-ray television. *In Femseh-undKino-Technik.* **33**(2), 41–2 (February1979).
75. Henderson, L. J. *The Fitness of the Environment.* Beacon Press, Boston (1913).
76. Hiemenz, P. C. and Lodge. T. P. *Polymer Chemistry* 2nd ed. Boca Raton: CRC Press (2007).
77. Holzbock, W. G. *Automatic Control.* Reinhold Publish Corporation, New York (1958).
78. Horowitz, I. M. *Synthesis of Feedback Systems.* Academic Press, New York (1963).
79. Hranilovic, S. *Modulation and Constrained Coding Techniques for Wireless Infrared Communication Channels*, Masters thesis, Dept. Elec. and Comp. Eng., Univ. of Toronto, Canada (1999).
80. Hubener, K. H. Conceptions for optimization in radiation therapy. *In Strahlentherapie.*, **154**(12), 858–60 (December, 1978).
81. Hurst, P. A Comparison of Two Approaches to Feedback Circuit Analysis. *TEEE Trans. Education*, **35(**3), 253–261 (August, 1992).
82. Ifrah, G. *The Universal History of Numbers.* John Wiley & Sons, Inc., New York, p.47, (2000).
83. Iga, K. *Surface Emitting Lasers*, Electronics and Communications in Japan, Part 2, **82(**10), 70–82 (1999).
84. Inderscience. Computer database compresses DNA sequences used in medical research. ScienceDaily. pp. 1–2, (July 9, 2010) (2009). Website: http://www.sciencedaily.com/releases/2009/11/09111120105.html
85. Infrared Data Association (lrDA), Serial Infrared Physical Layer Link Specification, Version 1.2, http://www.irda.org, (November, 1997)
86. Izawa, K. *Introduction to Automatic Control.* Elsevier Publishing Company, Amsterdam (1963).
87. J. ACS. Organic ternary data storage device developed. *Journal of the American Chemistry Society.* pp. 1–2 (September 3, 2010). Website: http://www.physorg.com/print190451302.html
88. James, H. M., Nicholes, N. B., and Phillips, R. S. *The Theory of Servomechanisms.* Boston Technical Publishers Inc., Lexington (1964).
89. Jern, N. K. *The generative grammar of the immune system.* Nobel Lecture, 211–225 (December 8, 1984).
90. Jesdanun, A. *Rebuilding the internet.* The Modesto Bee, pp. D-1 and D-3 (Thursday, April 19, 2007).
91. Jesperson, O. *The Philosophy of Grammar.* London: George Allen & Unwin, LTD (1924/1968).
92. Johnson, D. E. and Johnson, J. R. *Graph Theory: With Engineering Applications*, The Ronald Press Co., New York, p. 277 (1972). (b) Ibid. p. 278.
93. Johnson, D. E. and Johnson, J.R. *Graph Theory: With Engineering Applications.* The Ronald Press Company, New York (1972).
94. Kelly, R. D. Electronic Circuit Analysis and Design by Driving-Point Impedance Techniques, *IEEE Trans. on Education*, **El3**(3), 154–167 (September, 1970).

95. Khorramabadi, H., Tzeng, L., and Tarsia, M. A1.06Gb/s-31dBmtoOdBm BiCMOS Optical Pre amplifier Featuring Adaptive Transimpedance, *IEEE ISSCC Digest of Tech.* Papers, pp. 54–55 (February, 1995).
96. Kipiniak, W. *Dynamic Optimization and Control*. The MIT Press and John Wiley & Sons Inc., New York (1961).
97. Knuth, D. E. *The Art of Computer Programming: Volume 2 Semi numerical Algorithms*, Addison-Wesley Publishers, Reading, p.149 (1997).
98. Knuth, D. E. *The Art of Computer Programming: Volume 2 Semi numerical Programming*, Addison-Wesley Publishers, Reading, p. 169–170 (1997).
99. Koenig, H. E. and Blackwell, W. A. *Electromechanical System Theory*. McGraw-Hill Book Company Inc., New York (1961).
100. Kotz, S. and Johnson, N. I. *Encyclopedia of Statistical Sciences*, John Wiley & Sons, Inc., New York, p. 39 (1982).
101. Kotz, S. and Johnson, N. I. *Encyclopedia of Statistical Sciences*, John Wiley & Sons, New York (1982).
102. Krinsky, J. A. *Testing fiber optic data links for sensitivity to high gamma radiation dose rates*. Delivered as a paper at the Second Annual Military Fiber Optics Conference-West in 1988 and publishing in Boston: Information Gatekeepers (1988).
103. Ku, Y. H. *Analysis and Control of Linear Systems*. International Textbook Company Scranton (1962).
104. Ku, Y. H. Resume of Maxwell's and Kirchoff's rules for network analysis. *J. Frank. Inst.*, **253**, 211–224 (March, 1952).
105. Kuo, B. C. *Automatic Control Systems*. Prentice-Hall Inc., Englewood Cliffs (1962).
106. Ladu, M. and Randaccio, P. *A simple device for photographic dosimeter calibration. In Fisica e Tecnologia*, **3**(4), 276–80 (1980).
107. Lago, P. J. Open-loop stochastic control of pharmacokinetic systems: a new method for design of dosing regimens. *In Computers and Biomedical Research.* **25**(1), 85–100 (February, 1992).
108. Lagowski, J. J. *MacMillan Encyclopedia of Chemistry*. Volume 3 New York: MacMillan Simon & Schuster (1997).
109. Langley, L. L. *Homeostasis*. Reinhold Publishing Corporation, New York (1965).
110. Lenert, L. A., Lurie, J., Sheiner, L. B., Coleman, R., Klostermann, H., and Blaschke, T. F. Advanced computer programs for drug dosing that combine pharmacokinetic and symbolic modeling of patients. *In Computers and Biomedical Research.* **25**(1), 29–42 (February, 1992).
111. Levinson, S. A. and MacFate, R. P. *Clinical Laboratory Diagnosis*. Lea & Febiger, Philadelphia (1969).
112. Li, M. and Vitany, P. *An Introduction to Kolomogov Complexity and its Application*, Second Edition. Springer, New York (1997).
113. Li, M. and Vitanyi, P. Minimum Description Length Induction, Bayesiansim, and Kolmogorov Complexity. *IEEE Transactions on Information Theory*, **46**(2), (March 2000).
114. Li, M. and Vitanyi, P. *An Introduction to Ko1mogorov Complexity and Its Applications*, Springer, New York, p. 186 (1997).
115. Li, M. and Vitanyi, P. M. H. *An Introduction to Kolmogorov Complexity and its Applications*. Springer, New York (1993/1997).
116. Lindoff, D. P. *Theory of Sampled-Data Control System*. John Wiley & Sons Inc., New York (1965).
117. Lutter, abide. p. 374.

118. Lutter, L. C. Deoxyribonucleic acid. In McGraw-Hill Encyclopedia of science & technology. McGraw-HillPublishers, New York, pp. 373–379 (2007).
119. Lwoff, A. *Biological Order*. The MIT Press, Cambridge (1962).
120. Macmillan, R. H. *Progress in Control Engineering*, Academic Press, New York, pp. 4–5 (1964).
121. Macmillan, R. H. *Progress in Control Engineering*. Academic Press, New York (1964).
122. Macmillian, R. H., Higgins, T. J., and Naslin, P. *Progress in Control Engineering*, Academic Press, New York, p. 4 (1964).
123. Macmillian, R. H., Higgins, T. J., and Naslin, P. *Progress in Control Engineering*. Academic Press Inc. Publishers, New York (1964).
124. March 16, 2013, pp. 1-14. Website: http://rpi.edu/dept/bcbp/molbiochem/MBWeb/mbl/part2/sugar.hlmt
125. Martin-Lof, P. *Infor. And Contr.*, **9**(6), 602–619 (1966).
126. Martin-Lof, P. The definition of random sequences, *Info. &Cont.*, **9**, 602–619 (1966).
127. Martin-Lof, P. The definition of random sequences. *Information and Control*, **9**, 602–619 (1966).
128. Mason, S. J. Feedback theory: Some properties of signal flow graphs. *Proc. IRE*, (September, 1953).
129. Mason, S. J. and Zimmermann, H. J. *Electronic Circuits, Signals and Systems*, J. Wiley and Sons Inc., Chapter 5, New York (1960).
130. Mason, S. J. Feedback theory—some properties of signal flow graphs. *PROC. IRE*, **41**, 1144–1156 (September, 1953).
131. Mason, S. J. Feedback Theory- Some properties of signal flow graphs, *Proc. IRE*, **64**, 1144–1156 (September, 1953).
132. Menninger, K. *Number Words and Number Symbols: A Cultivated History of Numbers*. The MIT Press, Cambridge (1969).
133. Menshutkin, V. V., Suvorova, T. P., and Balonov, L. Y. Models of functional shutdowns of 1 hemisphere and neuro pharmacological effects on deep cerebral structures. In *Fiziologiya Cheloveka*. **7**(5), 880–888 (1981).
134. Meyer, R. G. and Mack, W. D. A Wideb and Low-Noise Variable-Gain BiCMOS Transimpedance Amplifier, *IEEE J. Solid-State Circuits*, **29**(6), 701–706 (June 1994).
135. Miller, G. M. *Modern Electronic Communication*. Prentice Hall Career & Technology, Englewood Cliffs (1993).
136. Mishkin, E. and Braun, L. *Adaptive Control Systems*. McGraw-Hill Book Company Inc., New York (1961).
137. MIT.What determines the length of words? MIT researchers say they know. *Phys.org.*, pp. 1–3, (February 18, 2011) (2013). Website: http://www.physorg.com/print216545475.html
138. Moles, A. *Information Theory and Esthetic Perception*. University of Illinois Press, Urbana (1966).
139. Monzon, J. E. and Guillen, S. G. Current defibrillator: New instrument of programmed current for research and clinical use. *In IEEE Transactions Bio-Medical Engineering*, **32**(11), 928–34 (1985).
140. Moore, J. A. McGraw-Hill Encyclopedia of Chemistry, McGraw-Hill Publishers, New York (1993).
141. Moreira, A. J. C., Valadas, R. T., and Oliveira Duarte, A. M. de. Optical interference produced by artificial light. *Wireless Networks*, **3**, 131–140 (1997).
142. Moreira, A. J. C., Valadas, R. T., and Oliveira Duarte, A. M. de. Characterization and Modelling of Artificial Light Interference in Optical Wireless Communication Systems. *IEEE Int. Symp. on Personal Indoor Mobile Radio Commun.*, **1**, 326–331, (September, 1995).

143. Muggleton, S. B. Exceeding human limits. *Nature*, **440**, 409–410 (March 23, 2006).
144. Muoi, T.V. Receiver Design for High-Speed Optical-Fiber Systems, *IEEE J. Lightwave Tech.*, **LT–2**(3), 243–267 (June 1984).
145. Murrill, P. W. *Automatic Control of Processes.* Scranton: International Textbook Company (1967).
146. Nagel, E. *Automatic Control*. Simon and Schuster, New York (1948).
147. Nakamura, M. and Ishihara, N. 1.2V, 35mW CMOS optical transceiver ICs for 50Mbitls burst-mode communication, *Electronic Letters*, **35**(5), 394–395 (March 4, 1999).
148. Narasimhan, R., Audeh, M. D., and Kahn, J. M. Effect of electronic-ballast fluorescent lighting on wireless infrared links, *lEE Proc. Optoelectronics*, **143**(6), 347–354 (December, 1996).
149. Neuteboom, H. et. Al. A single battery, 0.9 V operated digital sound processing IC including AD/ DA and TR receiver with 2 mW power consumption. *TEEE TSSCC Dig. Tech. Papers*, pp. 98–99 (February, 1997).
150. Newton, G. C. and Gould, L. A. *Analytical Design of Linear Feedback Controls*. John Wiley & Sons Inc., New York (1957).
151. Nikolic, B. and Matjanovic, S. A General Method of Feedback Amplifier Analysis. *IEEE Proc. Int.Symp. Circuits Systems*, **3**, 415–418 (May 1998).
152. Nuecherlein, J. E. and Weiser, P. J. *Digital Crossroads*. The MIT Press, Cambridge, p. 129 (2005).
153. Nuecherlein,J.E. and Weiser, P.J. *Digital Crossroads.Cambridge*, The MIT Press (2005).
154. Ochoa, A. A Systematic Approach to the Analysis of General and Feedback Circuits and Systems Using Signal Flow Graphs and Driving-Point Impedance, *TEEE Trans. Circ. and Syst. ll*, **45**(2), 187–195 (Feb, 1998).
155. Abide.
156. Oguztoreli, M. N. *Time-Lag Control Systems*. Academic Press, New York (1966).
157. Ohhata, K., Masuda, T., Imai, K., Takeyari, R., and Washio, K. A Wide-Dynamic Range, High-Tran simpedanceSi Bipolar Pre amplifier IC for 10-Gb/s Optical Fiber Links. *IEEE JSolid-State Circuits*, **34**(1), 18–24 (January, 1999).
158. Oldenbourg, R. C. and Sartorius, H. A Uniform Approach to the Optimum Adjustment of Control Loops in Frequency Response Symposium. *The American Society for Mechanical Engineers*, New York (1953).
159. Oliver, M. E. *Laboratory Manual to Modern Electronic Communication.* Regents/Pientice Hall Englewood Cliffs(1993).
160. Owen, B. PIN-GaAs FET optical receiver with a wide dynamic range. *Electronic Letters*, **18**(14), 626–627 (July, 8 1982).
161. Palmer, J. DNA computer calculates square roots. BBC News, pp. 1–3 (June 2, 2011). Website: http://www.bbcco.uk/news/science-environment-13626583?print=true
162. Palojarvi, P., Ruotsalainen, T., and Kostamovaara, J. A Variable Gain Transimpedance Amplifier Channel with a Timing Discriminator for a Time-of-Flight Laser Radar. *Proc. European Solid State Circuits Conf*, pp. 384–387 (September, 1997).
163. Panteleev, Q. A., Vaneeva, L. I., Demchenko, E. N., and Semenova, G. V. Problems of selecting the reagent concentration in the solid-phase immune enzyme method for determining antibody concentrations. *In Zhurnal Mikrobiologii Epidemiologii Immunobiologii.* **4**, 80–5, (April, 1987).
164. Penrose, R. *The Emperor's New Mind: Concerning Computers. Minds and The Laws of Physics*, Oxford, Oxford University Press (1989).
165. Personick, S. D. Receiver Design for Digital Fiber Optic Communications Systems, I and II, *BellSyst. Tech. Journal*, **52**(6), 843–886,(July–August 1973).

166. Personick, S. D. *Fiber Optics, Technology, and Applications*, Plennum Press, New York (1985).
167. Peschon, J. *Disciplines and Techniques of Systems Control*. Blaisdell Publishing Company, New York (1965).
168. Peslin, R., Saunier, C., Duvivier, C., and Marchand, M. Analysis of low-frequency lung impedance in rabbits with nonlinear models. *In the Journal of Applied Physiology*, **79**(3), 771–80 (September 1995).
169. Petri, C., Rocchi, S., and Vignoli, V. "High Dynamic CMOS preamplifiers for QW diodes." *Electronics Letters*, **34**(9), 877–878, (April 30, 1998).
170. Physical Review of Letters. Sequence compositional complexity of DNA through an entropic segmentation method. pp. 1–2, (October 15, 2010). Website: http://prl.aps.org/abstract/PRI/v80/i6/p1344_1
171. Plouffe, L. Biological modeling on a microcomputer using standard spreadsheet and equation solver programs: the hypothalamic-pituitary-ovarian axis as an example. *In Computers and Biomedical Research*. **25**(2), 117–30 (April, 1992).
172. Pollack, A. *DNA sequencing caught in deluge of data*. The New York Times. pp. 1-5. (December 1, 2011). Website: http://www.nytimes.com/2011/12/01/business/dna-sequencing-caught=in-deluge-of-data.html
173. Popov, E. P. *The Dynamics of Automatic Control Systems*. Pergamon Press Ltd, London (1962).
174. Porter, A. *Introduction to Servomechanisms*. Methuen & Company Ltd, London (1950).
175. Princeton University. *Redundant genetic instructions in "junk" DNA support healthy development*. Science News, pp. 1–3 (July 16, 2010). Website: http://esciencenews.com/articles/2010/07/16redundant-genetic-instructions-junk-dna-suppor
176. Principles of Frequency Response. Instrument Society of America, Pittsburgh (1958).
177. Qasaimeh, O., Singh, S., and Bhattacharya, P. Electro absorption and electro optic effect in SiGe-Si quan tum wells: realization of low-voltage optical modulators. *IEEE J Quantum Electronics*, **33**(9), 1532–1536 (September, 1997).
178. Qvarnstrom, B., Schutt, T., and Runnstrom-Reio, V. *Instruments and Measurements*. Academic Press, New York (1965).
179. R. H. Macmillan, T. J. Higgins, and P. Naslin (Eds.). *Progress in Control Engineering*. Academic Press Inc., Publishers, New York (1964).
180. Ranjbaran, E. *Modeling of Electrical Engineers*. Dissertation. University of Missouri, Columbia (1974).
181. Ranjbaran, S. E. State variable techniques in phannacokinetics using computer graphics. In IEEE Engineering Medical and Biological Society Annual Conference September 28–30, 1980 in Washington D.C., *IEEE*.
182. Richards, R. K. *Algorithmic Operations in Digital Computers*, D. Van Nostrand Company, Princeton, New Jersey, (1955).
183. Richards, R. K. *Arithmetic qperations in Digita Computers*, Van Nostrand Company, Inc, New York (1955).
184. Richards, R.K. *Arithmetic Operations in Digital Computers*, D. Van Nostrand Company, Inc., Princeton, p. 184 (1955).
185. Rissanen, J. Modeling by shortest data description. *Automatica*, **14**, 465–471 (1978).
186. Ritter, M. B., Gfeller, F., Hirt, W., Rogers, D., and Gowda, S. Circuit and System Challenges in IR Wireless Communication. *IEEE ISSCC Digest of Tech. Papers*, pp. 398–399 (February, 1996).
187. Rose, N. R. and Friedman, H. *Manual of Clinical Immunology*. American Society for Microbiology, Washington (1980).

188. Rosenstark, S. Feedback Amplifier Principles, Macmillan Publishing Co., Chapter 2, New York (1986).
189. RPI.edu (2013) "Carbohydrates-Sugars and Polysaccharides".
190. Rusinofl, S. E. *Automation in Practice*. American Technical Society, Chicago (1957).
191. Rybacki, J. J and Long, J. W. *The Essential Guide to Prescription Drugs*. Harper Perennial, New York (1996).
192. S. E. Miller and I. P. Kaminow (Eds.), *Optical Fiber Telecommunications II*, Academic Press, SanDiego (1988).
193. Sackinger, E., Ota, Y, Gabara, T. J., and Fischer, W. C. 15mW, 155Mb/s CMOS Burst-Mode Laser Driver with Automatic Power Control and End-of-Life Detection. *TEEE TSSCC Dig. Tech. Papers*, pp.386–387 (February, 1999).
194. Science Daily, pp. 1–3, (January 24, 2013). Website: http://www.sciencedaily.com/2013/01/130123133432.htm
195. Sebeok, T. A. Portraits of Linguists: A Biographical Source Book for the History of western Linguistics 1746-1963. Indianan University press, Bloomington (1966).
196. Sedra, A. S. and Smith, K. C. *Microelectronic Circuits*, 4th Ed., Oxford Univ. Press, Chapter 8, New York (1998).
197. Seifert, W. W. and Steeg, C. W. *Control Systems Engineering*. McGraw-Hill Book Company Inc., New York (1960).
198. Shannon, C. E. A mathematical theory of communication. *Bell Sys. Tech. Jour.*, **27**, 379–423 and 623–656 (1948).
199. Shannon, C. E. *An algebra for theoretical genetics*. Massachusetts Institute of Technology, Ph.D. Thesis (1940).
200. Shannon, C. E. and Weaver, W. *The Mathematical Theory of Communication*. University of Illinois Press, Urbana (1949).
201. Shannon, C. E. *Bell Labs. Tech. Jour.*, **27**, 379–423 and 623–656 (1948).
202. Shi, Y. et al., Low (sub-1-Volt) Half wave Voltage Polymeric Electro-optic Modulators Achieved by Controlling Chromophore Shape. *Science*, **288**, 119–122 (April 7, 2000).
203. Shinner, S. M. *Control System Design*, John Wiley & Sons Inc., New York (1964).
204. Shinners, S. M. *Control System Design*, John Wiley & Sons Inc., New York, p. 25 (1964). (b) Ibid. p.28.
205. Shinners, S. M. *Control System Design.* John Wiley & Sons Inc., New York (1964).
206. Smith, O. J. M. *Feedback Control Systems*. McGraw-Hill Book Company Inc., New York (1958).
207. Smithand, R. G. and Personick, S. D. Receiver design for optical fiber communication systems, in *Semiconductor Devices for Optical Communication*, 2nd Ed., H. Kresse (Ed.), Chapter 4, Springer-Verlag, Berlin, Germany, pp. 89–160 (1982).
208. Solodov, A. V. and Fuller, A. T. *Linear Automatic Control Systems with Varying Parameters*. American Elsevier Publishing Company Inc., New York (1966).
209. Solodov, A. V. *Linear Automatic Control Systems with Varying Parameters*. American Elsevier Publishing Company Inc., New York (1966).
210. Solodovnikov, V. V. *Statistical Dynamics of Linear Automatic Control Systems*. D. Van Nostrand Company Ltd. London (1965).
211. Solomonoff, R. *J. Inf. & Cont.*, **7**, 1–22 and 224–254 (1964), Kolmogorov, A. N. *Pro. Inf. & Trans.*, **1**, 1–7 (1965) and Chaitin, G. J. *Jour. ACM*, **16**, 145–159 (1969).
212. Stanford University Medical Center: Biological transistor enables computing within living cells. ScienceDaily, pp. 1–5, (March 9, 2013). Website: http://www.sciencedaily.com/releases/2013/03/130328142400.htm

213. Steiner, R. P. *Folk Medicine: The Art and the Science*, American Chemical Society: Washington, DC, p. 19 (1986). (b) Ibid. pp. 1–131 (c) Ibid. p. 116.
214. Stem, P. and Hines, P. J. Neuroscience: Systems-Level Brain Development. *Science*, **310**, 801 (November 4, 2005).
215. Steyaert, M., Crols, J., and Gogaert, S. Switched-Opamp, a Technique for Realizing Full CMOS Switched-Capacitor Filter sat Very Low Voltages. *Proc. Eur. Solid-State Circuits Conf.*, pp.178–181, (September 1993).
216. Stick, R. V. Carbohydrates: The Sweet Molecules of Life.New York: Academic Press (2001).
217. Stiny, G. and Gips, J. *Algorithmic Aesthetics: Computer Models for criticism and Design in the Arts*. Berkeley: University of California Press (1978).
218. Sugimoto, Y. and Kasahara, K. Future Prospects of VCSELs: Industrial View, Vertical Cavity Lasers, Tech. for a Global Information Infrastructure, 1997 Dig. *IEEEILEOS Summer Topical Meetings*, pp. 7–8, (August, 1997).
219. Svoboda, E. *The DNA transistor*. Scientific American, p. 46. (December 2010).
220. Swartz, R. G., Ota, Y., Tarsia, M. J., and Archer, V. D. A Burst Mode, Packet Receiver with Precision Reset and Automatic Dark Level Compensation for Optical Bus Communications, *Symp. on VLSI Technology*, Kyoto, Japan, pp. 67–68 (May 1993).
221. Tanabe, A., Soda, M., Nakahara, Y., Furukawa, A., Tamura, T., and Yoshida, K. A Single Chip 2.4 Gb/ s CMOS Optical Receiver IC with Low Substrate Crosstalk Pre amplifier. *IEEE ISSCC Digest of Tech.. Papers*, pp. 304–305 (February, 1998).
222. TGen. New technology reduces storage needs and costs for genomic data. ScienceDaily, pp. 1–3, (July 9, 2010). Website: http://www.sciencedaily.com/releases/2010/07/100706150614.html
223. Thaler, G. J. and Pastel, M. P. *Analysis and Design of Nonlinear Feedback Control Systems*. McGraw-Hill Book Company Inc., New York (1962)
224. Thaler, G. J. *Elements of Servomechanism Theorv*. McGraw-Hill Book Company Inc., New York (1955).
225. The Neuron Doctrine Redux. *Science*, **310**, 791–793 (November 4, 2005).
226. Tice, abide.
227. Tice, abide.
228. Tice, B. *Representing Environmental Data: Graphing & B1ock Diagram App1ications*. Paper presented as a poster at the 45th lEST ATMConference, Ontario, California USA, (May 5, 1999)
229. Tice, B. *Chromatic Aspects of the Signal Flow Diagram*. A poster session for the 78th Pacific Division of American Association for the Advancement of Science, Corvallis, Oregon (OR) (June 23, 1997).
230. Tice, B. *Interpersonal Perception: A Monograph. Psycho1ogy Monograph Series*. Cupertino, Advanced Human Design (1998).
231. Tice, B. S. A Theoretical Model of Feedback in Pharmacology Using the Signal Flow Diagram. Doctorial Dissertation in Chemistry. Published by U.M.I., *Ann Arbor, Michigan U.S.A* (1996).
232. Tice, B. S. "Feedback systems for nontraditional medicines: A case for the signal flow diagram". *Journal of Pharmaceutical Sciences*, **87**(11), 1282–1285 (1998).
233. Tice, B. S. A *Theoretical Model of Feedback in Pharmacology Using the Signal Flow Diagram*. Bloomington: 1stBooks Publishers (2001).
234. Tice, B. S. *A theory on neurological systems Part 1*. Technical Report-Advanced Human Design, **1**(1) (October, 2005).

235. Tice, B. S. A radix 4 based system for communications theory, Technical Paper, *Advanced Human Design*, **1**(2), 1–3 (December, 2006).
236. Tice, B. S. *Aspects of Kolomogorov: The Physics of Information*. Denmark, River Publishers (2009).
237. Tice, B. S. *The analysis of binary, ternary, and quaternary based systems for communications theory*. Poster for the SPIE Symposium on Optical Engineering and Application Conference, San Diego, California (August 10–14, 2008).
238. Tice, B. S. The Turing Machine: A Question of Linguistics?" A paper given at the Pacific Division of the American Association for the Advancement of Science [AAAS] annual meeting at Oregon State University, (June 22–26, 1997). A copy of the paper is reproduced in Appendix A of Thought Function and Form: *The Language of Physics*, B. S. Tice author, published by 1st Books Library, Indianapolis: 2004, pp. 207–214.
239. Tice, B. S. *The use of a radix 5 base for transmission and storage of information*, Poster for the Photonics West Conference, San Jose, California (Wednesday January 23, 2008).
240. Tice, B. S. *Two Models of Information*, 1st Books Publishers, Bloomington (2003).
241. Tice, B. S. (In press) *Formal Constraints to Formal Languages*. Author House, Indianapolis.
242. Tice, B. *Signal Flow Diagrams and Biotechnology*. A poster session for the 1997 NCASM Annual Spring Meeting, Santa Clara University, Santa Clara (April 5, 1997).
243. Tice, B. *The Multicolored Arrows: Chromatic Aspects of the Signal Flow Diagram*. Presented at the National American Chemical Society Meeting, Las Vegas, NV (September 7–11, 1997).
244. Tice, B. *The Use of Signal Flow Diagrams in Chemical Analysis*. A poster session for the 15th Annual American Peptide Symposium, Nashville, TN, (June 17, 1997).
245. Tice, B.S. Feedback systems for nontraditional medicines: A case for the signal flow diagram. *Journal of Pharmaceutical Sciences*, **87**, 1282–1285(Number 11, 1998).
246. Tice, B. S. *A Level of Martin-Lof Randomness*. Science Press, New Hampshire (2012).
247. Tice, B. S. *Language and Godel's Theorem*: A Revised Edition. River Publishers, Denmark (2013).
248. Tice, B. S. *Formal Constraints to Formal Languages*, Author House, Bloomington in press.
249. Tomovic, R. *Sensitivity Analysis of Dynamic Systems*. McGraw-Hill Book Company Inc., New York (1963).
250. Tou, J. T. *Modem Control Theory*. McGraw-Hill Book Company, San Francisco (1964).
251. Tsien, H. S. *Engineering Cybernetics*. McGraw-Hill Book Company Inc., New York (1954).
252. Tucker, G. K. and Willis, D. M. A *Simplified Technique of Control System Engineering*. Honeywell Regulatory Company, Philadelphia (1958).
253. Tufte, E. *Envisioning Information*. Graphics Press, Cheshire (1990).
254. Tufte, E. *The Visual Display of Quantitative Information*. Graphics Press, Cheshire (1983).
255. Tustin, A. Direct Current Machines for Control Systems. *The Macmillan Company*, pp. 45–46, New York (1952).
256. University of Hertfordshire. New method to measure the redundancy of information. ScienceDaily, pp. 1–3, (March 19, 2013). Website: Http://sciencedaily.com/releases/2013/02/130214132809.htm
257. University of Reading. Genetic inspiration could show the way to revolutionise information technology. pp. 1–2, (May 28, 2010). Website: http://www.physorg.com/print196961035.html

258. University of Wisconsin-Madison. Self-assembling polymer arrays improve data storage potential. Escience news, pp. 1–2 (September 3, 2010). Website: http://esciencenews.com/articles/2008/08/14/self assembling.polymer.arrays.improves.data.st.
259. Uspensky, V. A. An introduction to the theory of ko1mogorov complexity. O. Watanabe (Ed.). *Ko1mogorov Complexity and Computational Complexity* (Springer Verlag, Berlin), p. 87 (1992).
260. vanden Broeke, L. A. D. and Niewkerk, A. J. Wide-Band Integrated Optical Receiver with Improved Dynamic Range Using a Current Switch at the Input, *IEEE J. Solid-State Circuits*, **28**(7), 862–864 (July, 1993).
261. Vise, D. A. and Malseed, M. *The Google Story*. Delacorte Press, New York (2005).
262. Vitanyi, P. and Li, M. Minimum Description Length Induction. Bayyesnism and Kolmogorov Complexity. arXiv;cs/9901014vl [cs. LG] 27 Jan 1999, pp. 1–35 (2008).
263. Von Neumann, J. Probabilistic Logistics and the Synthesis of Reliable Organismsf rom Unreliable Components. John von Neumann Collected Works: Volume V. A. H. Taub (Ed.) The Macmillan Company, New York, pp. 329–378 (1956/1963).
264. Wallace, C. S. and Boulton, D. M. An Information Measure for Classification. *Comp. J.* **11**(2), pp. 185–195 (1968).
265. Warren, J. E. *Control Instrument Mechanisms Kansas City*, The Bobbs-Merrill Company Inc., (1967)
266. Wei, D., Shea, M., Saidel, G. M., and Jones, S. C. Validation of continuous thermal measurement of cerebral blood flow by arterial pressure change. *In Journal of Cerebral Blood Flow and Metabolism*, **13**(4), 693–701 (July, 1993).
267. West, J. C. *Analytical Techniques for Non-Linear Control Systems*. D. Van Nostrand Company Inc., New York (1960).
268. Weyl, H. Symmetry. p. 710, J. R In Newman (Ed.) *The World of Mathematics, Simon and Schuster*, New York, pp. 671–724 (1956).
269. Wikipedia "*Algorithmic Information Theory*". pp. 1–7, (May 2, 2013). Website: http://en.wikipedia.org/wiki/Algorithmic_information
270. Wikipedia "*Big Data*". pp. 1–8, (March 18, 2013). Website: ttp://en.wikipedia.org/wiki/Big_data
271. Wikipedia "*Chemical formula*". pp. 1–11, (March 16, 2013). Website: http://en.wikipedia.org/wiki/Chemical_formula
272. Wikipedia "*Data compression asymmetry*". pp. 1–2, (May 2, 2013). Website: http://en.wikipedia.org/wiki/Data_compression_asymmetry
273. Wikipedia "*Data Compression*". pp. 1–10, (May 2, 2013). Website: http://en.wikipedia.org/wiki/Compression_algorithm
274. Wikipedia "*Entropy Encoding*" (2013). Hrrp://en.wikipedia.org/wiki/Entropy_encoding
275. Wikipedia "*Fischer Projections*". pp. 1–3, (March 16, 2013). Website: http://en.wikipedia.org/wki/Fischer_projection
276. Wikipedia "*Huffman Coding*". pp. 1–10 (May 2, 2013). Website: http://en.wikipedia.org/wiki/Huffman_coding
277. Wikipedia "*Language of mathematics*". pp. 1–8 (March 23, 2013). Website: http://en.wikipedi.org/wiki/Mathematics-as-a-language
278. Wikipedia "*Polymer*". pp. 1–25 (May 2, 2013). Website: http://en.wikiedia.org/wiki/Polymers
279. Wikipedia "*QWERTY*". pp. 1–15 (March 16, 2013). Website: http://en.wikipedia.org/wiki/QWERTY
280. Wikipedia "*Simplified molecular-input line-entry system*". pp. 1–12 (March 16, 2013) Website: http://en.wikipedia.org/wiki/Simplified_molecular-input_line-ent...

281. Wikipedia "*Smiles arbitrary target specification*". pp. 1–7 (March 16, 2013). Website: http://en.wikipedia.org/wiki/Smiles_arbitrary_target_specification
282. Wikipedia "*Systems theory*". pp. 1–13 (March 18, 2013). Website: http://en.wikipedia/wiki/General_systems_theory
283. Wikipedia "*Text messaging*". pp. 1–21 (March 16, 2013). Website: http://en.wikipedia.org/wiki/Texting
284. Wikipedia "*Universal Code (data compression)*". pp. 1–3 (May 2, 2013). Website: http://en.wikipedia.org/wiki/Universal_coe_(data_compression)
285. Wikipedia "*Copolymer*". Wikipedia pp. 1–5 (September 4, 2010). Website: http://en.wikipedia.org/wiki/Copolymer.
286. Wikipedia "*Polymer*". Wikipedia, p. 1. (September 4, 2010). Website: http://en.wikipdia.org/wiki/Polymer.
287. Wikipedia "*Synthetic genomics*". Wikipedia p. 1 (September4, 2010). Website: http://en.wikipeida/wiki/Synthetic-genomic.
288. Wikipedia. "*Minimum Description Length*". Wikipedia Encyclopedia, pp. 1–6 (2012). Website: http://en.wikipedia.org/wiki/Minimum description-length
289. Wilsonand, B. and Drew, J. D. Novel Transimpedance Amplifier Formulation Exhibiting Gain-Bandwidth Independence. *IEEE Proc. Tnt. Symp. Circuitsand Systems*, I, 169–172, (June, 1997).
290. Woerd, A. C. van der. Low-voltage low-power infra-red receiver for hearing aids, Electronics Letters, **28**(4), 396–398 (February 13, 1992).
291. Wright, R. *Three Scientists and Their Gods*. Times Books, New York (1988).
292. Yamazakiet, D. 156Mbit/s pre amplifier IC with wide dynamic range for ATM-PON application, *Electronic Letters*, **33**(15), 1308–1309 (July 7, 1997).
293. Yang, G. M., Macdougal, M. H., and Dapkus, P. D. Ultralow threshold vertical cavity surface emitting lasers obtained with selective oxidation, *Electronic Letters*, **31**(11), 886–888 (May 25, 1995).
294. You, F., Embabi, S. H. K., and Sanchez-Sinencio, E. Multistage Amplifier Topologies with Nested Gm-C Compensation, *IEEE J. Solid-State Circuits*, **32**(12), 2000–2011 (December, 1997).

INDEX

G

H

I

J

K

L

M

N

O

P